A History of Science in Society

A History of Science in Society

Volume II:
From the Scientific Revolution to the Present

Fourth Edition

Andrew Ede and Lesley B. Cormack

UNIVERSITY OF TORONTO PRESS
Toronto Buffalo London

© University of Toronto Press 2022
Toronto Buffalo London
utorontopress.com
Printed in the U.S.A.

ISBN 978-1-4875-2466-1 (paper)   ISBN 978-1-4875-3523-0 (EPUB)
                                      ISBN 978-1-4875-3522-3 (PDF)

Library and Archives Canada Cataloguing in Publication

Title: A history of science in society / Andrew Ede and Lesley B. Cormack.
Names: Ede, Andrew, author. | Cormack, Lesley B., 1957–, author.
Description: Fourth edition. | Includes bibliographical references and indexes. | Contents: Volume II: From the scientific revolution to the present.
Identifiers: Canadiana (print) 20210306785 | Canadiana (ebook) 20210306793 | ISBN 9781487524661 (v. 2 ; softcover) | ISBN 9781487535230 (v. 2 ; EPUB) | ISBN 9781487535223 (v. 2 ; PDF)
Subjects: LCSH: Science – History. | LCSH: Science – Social aspects – History.
Classification: LCC Q125 .E33 2022 | DDC 509 – dc2

We welcome comments and suggestions regarding any aspect of our publications – please feel free to contact us at news@utorontopress.com or visit us at utorontopress.com.

Every effort has been made to contact copyright holders; in the event of an error or omission, please notify the publisher.

We wish to acknowledge the land on which the University of Toronto Press operates. This land is the traditional territory of the Wendat, the Anishnaabeg, the Haudenosaunee, the Métis, and the Mississaugas of the Credit First Nation.

University of Toronto Press acknowledges the financial support of the Government of Canada and the Ontario Arts Council, an agency of the Government of Ontario, for its publishing activities.

ONTARIO ARTS COUNCIL
CONSEIL DES ARTS DE L'ONTARIO
an Ontario government agency
un organisme du gouvernement de l'Ontario

Funded by the   Financé par le
Government   gouvernement   Canada
of Canada   du Canada

CONTENTS

ILLUSTRATIONS

CONNECTIONS BOXES

ACKNOWLEDGMENTS

To Graham and Quin, who have grown up with this book – and who put up with two authors working in the house at the same time. We would also like to thank those people who helped make this book possible: our editor and publisher, friends and colleagues who read early drafts and gave advice, reviewers and users who have offered helpful criticism and forced us to defend our position, and all the amazing historians of science on whose shoulders (or toes) we stand.

INTRODUCTION

S cience has transformed human history. It has changed how we see the universe, how we interact with nature and each other, and how we live our lives. It may, in the future, even change what it means to be human. The history of such a powerful force deserves a full and multifaceted examination. Yet a history of science is unlike a history of monarchs, generals, steam engines, or wars because science isn't a person, an object, or an event. It is an idea, the idea that humans can understand the physical world.

This is a history of what happens when a legion of thinkers, at different times and from different backgrounds, turned their minds and hands to the investigation of nature. In the process, they transformed the world.

The history of science is such a vast subject that no single book about it can really be comprehensive, and so the story we tell examines science from a particular point of view. Some histories of science have focused on the intellectual development of ideas, while others have traced the course of particular subjects such as astronomy or physics. In this book, we have chosen to look at science from two related perspectives that we believe offer a window onto the historical processes that shaped the study of nature. First, we have examined the link between the philosophical pursuit of knowledge and the desire of both the researchers and their supporters to make that knowledge useful. There has always been a tension between the intellectual aspects of science and the application of scientific knowledge. The ancient Greek philosophers struggled with this problem, and it is still being debated today. The call in every age by philosophers and scientists for more support for "research for its own sake" is indicative of the tension between the search for knowledge and the pressure to apply that knowledge. What counts as useful knowledge differed from patron to patron and society to society, so that the Grand Duke Cosimo de' Medici and the United States Department of Energy looked for quite different "products" to be created by their clients, but both traded support for the potential of utility.

The tension between intellectual pursuits and demands for some kind of product not only was felt by many natural philosophers and scientists but has also led to controversy among historians of science. Where does science end and technology begin? they have asked. Perhaps the most famous articulation of this is the "scholar and craftsman debate." Historians of science have tried to understand the relationship between those people primarily interested in the utility of knowledge (the craftsmen) and those interested in the intellectual understanding of the world (the scholars). Some historians have denied the connection, but we feel it is integral to the pursuit of natural knowledge. The geographers of the early modern period provide a good example of the necessity of this interconnection. They brought the skills of the navigator together with the abstract knowledge of the mathematician. Translating the spherical Earth onto flat maps was an intellectual challenge, while tramping to the four corners of the globe to take measurements was an extreme physical challenge. Getting theory and practice right could mean the difference between profit or loss, or even life and death.

Our second aim has been to trace the history of science by its social place. Science does not exist in disembodied minds, but is part of living, breathing society. It is embedded in institutions such as schools, princely courts, government departments, and even in the training of soldiers. As such, we have tried to relate scientific work to the society in which it took place, tracing the interplay of social interest with personal interest. This has guided our areas of emphasis so that, for example, we give alchemy a greater allocation of space than some other histories of science because it was more socially significant than topics such as astronomy or physics in the same period. There were far more alchemists than astronomers, and they came from all ranks and classes of people, from peasants to popes. In the longer term, the transformation of alchemy into chemistry had a very great impact on the quality of everyday life. This is not to say that we neglect astronomy or physics, but rather that we have tried to focus on what was important to the people of the era and to avoid projecting the importance of later work on earlier ages.

In each chapter, we have highlighted at least one aspect of this interaction of science and society, from politics and religion to economics and warfare, under the heading "Connections." While each of these vignettes is part of the larger narrative of the book, they can also be read as individual case studies.

It is from the two perspectives of utility and social place that our subtitle comes. As we began to look at the work of natural philosophers and scientists over more than 2,000 years, we found ourselves more and more struck by the consistency of the issue of the utility of knowledge. Plato disdained the utility of knowledge, but he promoted an understanding of geometry. Eratosthenes used geometry to measure the diameter of the Earth, which had many practical applications. In the modern era, we have seen many cases of scientific work unexpectedly turned into consumer goods. The cathode ray tube, for instance, was a device created to study the nature of matter, but it ended up in the heart of the modern television. Philosophers and scientists have always walked a fine line between the role of intellectual and the role of technician. Too far to the technical side and a person will appear to be an artisan and lose their status as an intellectual. Too far to the intellectual side, a person will have trouble finding support because they have little to offer potential patrons.

Although the tension over philosophy and utility has always existed for the community of researchers, we did not subtitle our book "Philosophy and Utility." This is because the internal tension was not the only aspect of philosophy and utility that we saw over time. Natural philosophy started as an esoteric subject studied by a small, often very elite, group of people. Their work was intellectually important but had limited impact on the wider society. Over time, the number of people interested in natural philosophy grew, and as the community grew, so too did the claims of researchers that what they were doing would benefit society. Through the early modern and modern eras, scientists increasingly promoted their work on the basis of its potential utility, whether as a cure for cancer or as a better way to cook food. And, in large part, the utility of science has been graphically demonstrated in everything from the production of color-fast dyes to the destruction of whole cities with a single bomb. We have come to expect science to produce things we can use, and, further, we need scientifically trained people to keep our complex systems working – everything from testing the purity of our drinking water to teaching science in school. Our subtitle reflects the changing social expectation of science.

We have also made some choices about material based on the need for brevity. This book could not include all historical aspects of all topics in science or even introduce all the disciplines in science. We picked examples that illustrate key events and ideas rather than give exhaustive detail. For instance, the limited amount of medical history we include looks primarily at examples from medicine that treated the body as an object of research and thus as part of a larger intellectual movement in natural philosophy. We also chose to focus primarily on Western developments in natural philosophy and science, although we tried to acknowledge that natural philosophy existed in other places as well and that Western science did not develop in isolation. Especially in the early periods, Western thinkers were absorbing ideas, materials, and information from a wide variety of sources. By the seventeenth and eighteenth centuries, Western scholars were interacting with other cultures and exchanging information, although not on an equal footing. In later periods, Western science became a powerful tool for modernization and internationalization of countries around the world. Recently, scholars have begun to examine the interactions between Indigenous and traditional worldviews and knowledge and Western science, which we discuss briefly. *A History of Science* tells a particular – and important – story about the development of this powerful part of human culture, which has and continues to transform all our lives. To study the history of science is to study one of the great threads in the cloth of human history.

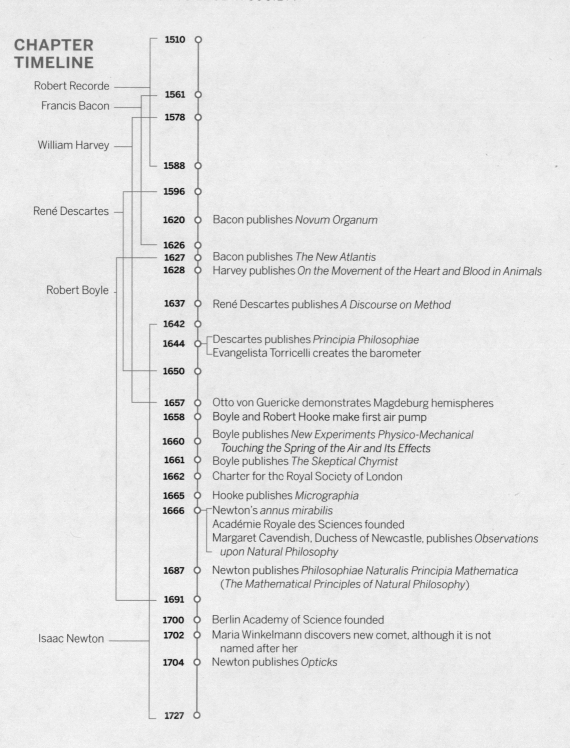

CHAPTER TIMELINE

Robert Recorde

Francis Bacon

William Harvey

René Descartes

Robert Boyle

Isaac Newton

1510	
1561	
1578	
1588	
1596	
1620	Bacon publishes *Novum Organum*
1626	
1627	Bacon publishes *The New Atlantis*
1628	Harvey publishes *On the Movement of the Heart and Blood in Animals*
1637	René Descartes publishes *A Discourse on Method*
1642	
1644	Descartes publishes *Principia Philosophiae* Evangelista Torricelli creates the barometer
1650	
1657	Otto von Guericke demonstrates Magdeburg hemispheres
1658	Boyle and Robert Hooke make first air pump
1660	Boyle publishes *New Experiments Physico-Mechanical Touching the Spring of the Air and Its Effects*
1661	Boyle publishes *The Skeptical Chymist*
1662	Charter for the Royal Society of London
1665	Hooke publishes *Micrographia*
1666	Newton's *annus mirabilis* Académie Royale des Sciences founded Margaret Cavendish, Duchess of Newcastle, publishes *Observations upon Natural Philosophy*
1687	Newton publishes *Philosophiae Naturalis Principia Mathematica* (*The Mathematical Principles of Natural Philosophy*)
1691	
1700	Berlin Academy of Science founded
1702	Maria Winkelmann discovers new comet, although it is not named after her
1704	Newton publishes *Opticks*
1727	

THE SCIENTIFIC REVOLUTION: CONTESTED TERRITORY

5

From 1543 to 1687 some of the giants of science lived, contemplated the natural world, and produced the underpinnings of modern science. Given the accomplishments of the period, it is no wonder that historians have agonized about the idea of an era of scientific revolution. Indeed, the twentieth-century discipline of the history of science began by focusing on the problem of the origin of modern science. The work of some of the discipline's founders concentrated on what this important transformation was and how it came to take place. In recent years, historians have begun to question whether such a revolution happened at all. Obviously, the answer depends on how revolution and science are defined: whether there was a gradual transformation of ideas, a gestalt switch, or a sociological innovation. We argue that there was a transformation in the investigation of the natural world, in which new ideas, methods, actors, aims, and ideologies vied with one another for a newly secularized role in the developing nation-states. This, indeed, was a scientific revolution.

The scientific revolution can be understood as a series of overlapping innovations, all important in the creation of modern science. First, natural philosophers took up the epistemological challenges of the ancients and developed a new methodology for uncovering the truth about the natural world. Second, in many different areas, but particularly in physics, astronomy, and mathematics, new theoretical models of the universe were developed. Further, those interested in the investigation of nature formed new institutions and organizations that began to perform the now largely secular tasks of evaluating scientific fact and determining who could be a natural philosopher. Finally, and perhaps most significantly, men interested in the underlying truths of nature developed a new ideology of utility and exploitation, a new structure for scientific practice, and a gentlemanly coterie of scientists who applied their social standards of behavior to the ideology of modern science.

The New Scientific Method: Francis Bacon and René Descartes

The rediscoveries of ancient natural philosophers and the challenges to that ancient knowledge in the sixteenth century caused scholars in the sixteenth and seventeenth centuries to turn to the epistemological question of how to determine truth from falsity. Perhaps spurred on by the religious turmoil of the sixteenth century, philosophers began to ask the question still fundamental to us today: How can we know what is true? This led to the development of a new form of scientific inquiry – a new "scientific method" – and a new way of articulating this search for certainty. In England this methodology was most fully articulated in the writing of Sir Francis Bacon (1561–1626). Bacon, although not himself a natural philosopher, proposed a reform of natural philosophy in the *Novum Organum* (1620) and *The New Atlantis* (1627). This program of reform was part of a grander scheme to transform all knowledge, especially legal knowledge and moral philosophy. Bacon believed that all human knowledge was flawed because of the Idols that all men carried with them. The Idols were the prejudices and preconceived ideas through which human beings observed the world. Bacon felt that the only way for natural philosophers to disabuse themselves of these Idols was to look at small, discrete bits of nature. The only way to be certain one understood these small bits was to study them in a controlled setting, isolated from the larger (uncontrolled) environment. Using this assumption, he introduced what has come to be called the inductive method. He suggested that increments of information could be gathered by armies of investigators, put together in tabular form, and explained by an elite cadre of interpreters.

Bacon described this in a section of *The New Atlantis* known as "Solomon's House." His methodology, although it appeared to be more democratic than earlier scholastic methods, proposed a means of controlling truth and knowledge by a small elite group who determined what could be studied and what answers were acceptable. In this attitude, he was probably influenced by the fact that he was trained as a lawyer and that he spent much of his political career as an advisor to Elizabeth I and then as Lord Chancellor for James I. He was thus accustomed to the idea of testing evidence in the public venue of a court. As the person most concerned with treason and heresy, he had a distrust of free-thinking and believed that ideas should be controlled by those whose position as custodians of the peace and security of the commonwealth best assured their credibility. One aspect of Bacon's job as Lord Chancellor included overseeing the use of torture in an age when evidence obtained by torture was considered reliable. For Bacon, knowledge was power, and thus an understanding of nature was important precisely because of the practical applications such knowledge would have. In many ways Bacon was a courtly philosopher, so the rhetoric of utility so well employed by Galileo was also present in his work.

This methodology was challenged on the continent by René Descartes and his followers, who preferred a deductive style based on skepticism. Descartes, like Bacon, came from an influential family and was trained as a lawyer. Unlike Bacon, he worked as a mathematics teacher and practitioner, rather than as a courtier, although in the end the temptation for patronage overcame the struggle involved in living by one's wits. In 1649 Descartes, at the age of 53 the most famous philosopher of his time, accepted the post of Court Philosopher to Queen Christina of Sweden. This was a lucrative post that, like the patronage of the Medicis for Galileo, offered both financial support and status in exchange for glorifying Christina's court and providing philosophical services. Unfortunately for Descartes, whose health was poor, Christina's idea of using his philosophical services was to have him call on her three times a week at five in the morning to instruct her. He was dead of pneumonia before the winter was over. The Swedes sent his body back to France but kept his head. This led to low-grade tension between France and Sweden for close to 200 years, until in 1809 the Swedish chemist Berzelius somehow managed to get Descartes's skull and return it to the French scientist Cuvier, who reunited it with the body.

In *A Discourse on Method* (1637) Descartes offered the first early modern alternative to Aristotle's epistemological system: his method of skepticism. He began by doubting everything, peeling away all layers of knowledge until he came to the one thing he knew was true: that as a thinking, doubting being, he must exist in order to think the doubting thought. He encapsulated this idea in the famous declaration *Cogito ergo sum: I think, therefore I am.* From this starting point he developed through deduction from first principles a series of universal truths that he knew to be self-evident. This deductive method owed its origins to geometric proof, which starts with a small set of sure premises or axioms and proceeds to more complex conditions. Interestingly, although Descartes used this mathematical model and developed new mathematical methodology, most of his scientific theories were explicitly nonmathematical. Also, he was not interested in using experimentation as a means to discover knowledge about nature. Since the senses could be fooled, right reasoning was a much more reliable arbiter in natural philosophical debate than any crude experiment or demonstration might be.

Both Bacon and Descartes attempted to find ways of reasoning that would produce certain knowledge in an age where certainty was giving way to probability. Bacon answered the question of how we can know what is true in a careful, conservative way, involving a hierarchy of knowledge made to appear as a democratic republic of scholars. Descartes answered in an individualistic anti-communal way, which gave more power to individual thinkers but did not, in the final analysis, create a community of scholars.

Mathematics as the Language of Natural Philosophy

One result of this new methodology with its debate about reliable and sure knowledge was the gradual ascendancy of mathematics as the language of natural philosophy. Galileo had been convinced that God created the world in number, measure, and weight, and many other scholars interested in nature echoed this sentiment. For Aristotle, weighing or measuring a substance did not tell you anything interesting about it, but for those investigating nature in the sixteenth and seventeenth centuries, knowing how heavy something was or how fast something went was surer knowledge than searching for final causes. They claimed that they only measured and observed rather than imposed underlying hypotheses they sought to prove. Sir Isaac Newton, for example, famously said *Hypotheses non fingo*: I feign no hypotheses. Following Galileo and Newton, natural philosophers increasingly looked for certainty through measurement rather than the analysis of cause.

Significant mathematical developments included the rediscovery of algebra and the development of the calculus; in addition, new and easier notation systems were devised. For example, Descartes instituted the use of a, b, and c for known variables and x, y, and z for unknowns. Mathematicians used these systems as they vied with one another for solutions to increasingly complex algebraic equations.

The creation of analytic geometry, primarily by Descartes but also by François Viète (1540–1603) and Pierre de Fermat (1601–65), placed a powerful tool in the hands of mathematicians and natural philosophers. By combining geometry and algebra, it became possible to transform geometric objects into equations and vice versa. This also opened the door to mathematizing nature, as everything from the trajectory of a cannonball to the shape of a leaf could be turned into a mathematical expression.

In turn, mathematicians began to look for new ways to measure areas under curves and to describe dynamic situations. The result was the invention of the calculus, independently developed by Isaac Newton in England and Gottfried Wilhelm Leibniz (1646–1716) on the continent. The calculus added infinitely small areas together under a curve or described the shape of the curve in formulaic terms. This allowed natural philosophers to accurately describe dynamic situations such as velocity and the motion of acceleration, something not possible with the older geometric and algebraic systems. Some mathematicians were concerned with the philosophical implications of the calculus, since it produced finite answers from the addition of infinite quantities, and the infinitely small and the infinitely large could, paradoxically, be equal. However, in an age that increasingly looked to the practical applications and utility of mathematics, the calculus proved to be an extremely fruitful device and was quickly taken up by the scientific community.

Leibniz was an influential German polymath, trained as a lawyer and employed most of his life by several German princes, especially the three dukes of Hanover. Unfortunately, his relationship with Duke Georg Ludwig (1660–1727) deteriorated, and when Ludwig became King George I of England, Leibniz was forbidden to enter the country. He was very critical of both Descartes's and Newton's work and became embroiled in an acrimonious dispute with Newton over who had invented the calculus first. This dispute, carried on by Leibniz in Germany and Samuel Clarke (1675–1729) in England, was one of the most famous philosophical disputes of the seventeenth and eighteenth centuries, as it ranged over issues of Newtonian natural philosophy and theology. In the end the participants died, and the matter was never resolved. It seems clear now that Newton and Leibniz did develop the calculus independently, with completely different notation and mathematical bases (Newton's was geometrical, while Leibniz's was analytical). In a sense Leibniz won, since the notation that has been used from the eighteenth century to the present is his rather than Newton's.

Mathematical Practitioners

One of the reasons mathematics became such a powerful tool in seventeenth-century natural philosophy was the presence of a new category of scientifically inclined men: the mathematical practitioners. Mathematics had been a quite separate area of investigation, and those interested in its issues had usually tied their studies to practical applications such as artillery, fortification, navigation, and surveying. In the early modern period, these mathematical practitioners provided the necessary impetus in the transformation of nature studies to include measurement, experiment, and utility. Their growing importance was the result of changing economic structures, developing technologies, and new politicized intellectual spaces such as courts, thus relating changes in science to the development of mercantilism and the nation-state. Mathematical practitioners claimed the utility of their knowledge, a rhetorical move that encouraged those seeking such information to regard it as useful.

Mathematical practitioners professed an expertise in a variety of areas. For example, Galileo's early work on physics and the telescope were successful attempts to gain patronage by using mathematics. Descartes advertised his abilities to teach mathematics and physics. Simon Stevin (1548–1620) claimed the status of a mathematical practitioner, including an expertise in navigation and surveying. William Gilbert (1544–1603) argued that his larger philosophical arguments about the magnetic composition of the Earth had practical applications for navigation. Leibniz used his mathematical power to act as an advisor on engineering projects for the dukes of Hanover. As well, many practitioners, including Thomas Hood (fl. 1582–98) and Edward Wright (1558–1615), demonstrated an interest in mapping and navigation.

This new interest in mathematics and in quantifying the behavior of the world sparked interest in probability. Mathematicians did not believe the world was capricious but that our incomplete knowledge of it limited our comprehension. The introduction of a mathematical evaluation of probability was a step toward understanding complex systems in which not all the determining factors could be known with certainty. Blaise Pascal (1623–62), Pierre de Fermat, and Christiaan Huygens (1629–95) all investigated the mathematical basis of prediction of games of chance, which were popular pastimes in the seventeenth century. Pascal's interest in the geometry of chance had wider implications than gambling at cards and dice, since it led him to develop his probabilistic argument for belief in God, now known as Pascal's Wager. He concluded that, although one cannot know with complete certainty if God exists, by using four possible conditions, one's best probable outcome would result from belief in God. If God did not exist, one lost nothing by believing in Him, but if He did exist, and one believed, one would be saved. Conversely, one lost a great deal by not believing in God if He did exist, while gaining nothing if He didn't exist. (See figure 5.1.)

5.1 PASCAL'S WAGER

	Do Not Believe in God	Do Believe in God
GOD DOES NOT EXIST	Nothing gained or lost	Nothing gained or lost
GOD DOES EXIST	Damnation	Salvation

By the end of the century Jacob Bernoulli (1654–1705) had codified the mathematics of probability, arguing that mathematics gave us the greatest certainty possible in an uncertain world. The concept of probability was not well accepted in physics, however, where Newton's universal laws seemed to provide certain, rather than probabilistic, answers. The shift of the foundation of physics from certainty to probability was one of the most traumatic transitions in modern science, but it would not happen for almost 200 years.

New Models of the Universe

All these new attempts to find a path to certain knowledge were crucial, since natural philosophers were making some radical suggestions about the makeup of the cosmos. The scientific revolution is most clearly identified with the development of a heliocentric model of the universe. This began with Copernicus, who claimed that the Earth revolved around the Sun and who developed a mathematical model to explain the movement of the planets. Natural philosophers had been slow to accept Copernicus's theory because it lacked a proper physical justification, such as Aristotle had provided through the concept of natural motion for his cosmological schema. Thus, the so-called Copernican Revolution (historians love to label innovations as revolutions) was incomplete until Sir Isaac Newton (1643–1727), the great English astronomer and mathematician, devised a mathematical model of motion that explained heavenly and earthly movement in a single physical system based on his concept of universal gravitation.

Isaac Newton: The Great Polymath

Isaac Newton was born January 4, 1643. His father died before he was born, and his early life was not happy, spent in conflict with a stepfather he disliked and a mother who expected him to run the family farm. Newton had no

CONNECTIONS

Science and the Marketplace: Mathematics for Sale

During the sixteenth century, many mathematical practitioners made their living by giving lectures and instruction, selling instruments, and performing a variety of mathematically based activities such as surveying or casting horoscopes. The booming metropolis of London provided an excellent marketplace for these skilled men, whose work demonstrated the interconnection of natural philosophy, mathematics, and the mercantile society of the time.

The first mathematical practitioner in London was Robert Recorde (1510–88). Trained at Oxford, he was largely responsible for introducing arithmetic and mathematics to a wider audience in England and for re-establishing a mathematical language and discipline into English scholarship. Recorde wrote a number of foundational books on mathematics in English. He was commissioned by the Muscovy Company (a merchant company that traded in northern Europe and was seeking the Northeast Passage to Cathay) to give lectures and to write a series of books on geometry, spherical geometry, astronomy, and navigation for use by their navigators. At the same time, he introduced Euclidian mathematics and algebra to an English audience, allowing natural philosophers to use mathematics in ways they had not done before.

Recorde was soon followed by other mathematical lecturers in London, sponsored by guilds, companies, or the City of London itself. Thomas Hood (1556?–1620), for example, was the first mathematics

aptitude for farming, and his mother despaired of finding him a livelihood. Fortunately, the local vicar noticed his scholarly potential and helped procure him a scholarship to Trinity College, Cambridge. Newton was a relatively undistinguished scholar, except for mathematics; he taught himself geometry from Descartes's works and algebra from Viète's. In due course, he was made a Fellow of Trinity College in 1664 and was appointed Lucasian Professor of Mathematics in 1669. The latter was a prestigious position, although he often lectured to empty rooms. His teacher, Isaac Barrow (1630–77), had to pull strings in order to get him this position, since Newton did not believe in the special divinity of Christ or in the Trinity. This made him a potential heretic, since he would not take the required oath of uniformity to the Church of England, which was normally required for any high academic or government lecturer paid by the City of London. In 1588, Hood petitioned William Cecil, Lord Burghley, to support a mathematics lectureship in London, to educate the "Capitanes of the trained bandes in the Citie of London."* Hood identified himself on the title pages of all his books until 1596 as Mathematical Lecturer to the City of London, sometimes advising interested readers to come to his house in Abchurch Lane for further instruction, or to buy his instruments. His books explained the use of mathematical instruments such as globes, the cross staff, and the sector, suggesting that his lectures and personal instruction would have emphasized this sort of instrumental mathematical knowledge and understanding. He was also known to have cast horoscopes, another way to make money.

Those who gave and attended these mathematical lectures had some expectation that they would be able to buy, sell, and use the instruments that were being discussed there. It is no surprise, therefore, that just as these lectures were being presented to the London community, mathematical instrument makers were beginning to ply their trade in increasing numbers in late sixteenth-century London. Thomas Gemini, a goldsmith, was probably the first English instrument maker starting in the 1550s, followed by Humphry Cole in the 1580s. After that, many men set up shop and sold navigational instruments, surveying instruments, maps, globes, and astrolabes, and provided the informal instruction for those who were interested in using them.

By 1610, there was a strong practical mathematical community living in London. A number of mathematical lectures had been sponsored, attended by a variety of audiences. Books and individual lessons explaining the use of mathematics and mathematical instruments had been produced, all leading to an increasing number of men trained in and sensitive to mathematical tools and explanations. A variety of men met in the instrument shops and at the mathematical lectures – gentlemen, scholars, merchants, and navigators. Mathematics was becoming both a language of commerce and of natural philosophy.

....

* British Library manuscript, Landsdowne 101, f. 56.

position. While Newton would not compromise his religious beliefs, he kept his views very private for his entire life.

In 1665 the Great Plague returned to England. It swept through Cambridge, and Newton was forced to return to his mother's home. The enforced isolation allowed him the opportunity to put together a number of ideas he had been developing through intensive reading during the past four years. Although it is unlikely that an apple really fell on his head, during the following year, his *annus mirabilis* (miraculous year), he worked out theories about gravitation, physics, and astronomy. As if that were not enough, he also created the calculus and began his investigations into optics and theories of light.

Although Newton studied planetary motion during that year, he did not publish his results until 22 years later. He was dissatisfied with his mathematical results to the question of why the orbits of the planets were circular, or nearly circular, and so put them away for a time in order to concentrate on alchemy and theology, including his long-standing interest in the books of Revelation and Daniel. He spent many hours over the next 13 years reading commentaries on scripture, studying the Bible, and constructing his personal theology. This theology was complex, most closely resembling an extreme form of Unitarianism or Arianism, which argued that Christ was not divine but the highest of God's created beings. Over his lifetime, Newton spent more time studying theology than any other subject.

In 1679 Robert Hooke (1635–1703), corresponding secretary of the Royal Society of London, wrote to Newton to find out what he was doing. It was Hooke's job to act as a kind of intellectual pen pal, communicating with members of the Royal Society and putting people with similar interests in touch. Newton and Hooke had a strained relationship, however, since Hooke had criticized Newton's earlier optical work. What, Hooke asked in 1679, would be the path of a body (a rock, for example) released from a high tower down to a rotating Earth? Newton replied that he was not presently engaged in natural philosophical investigations, but he suggested that it would spiral east to the center of the Earth because the angular velocity from the top of the tower was greater than on the Earth's surface. Hooke argued that this was not so: the path would be the result of a horizontal linear motion at constant speed combined with an attractive force toward the center and varied inversely with the square of the distance between the body and the Earth. This caused Newton to wonder whether his original question in 1666 as to why planetary orbits were circular was misdirected.

Newton set to work on mathematical models of elliptical orbits. In 1684 his friend Edmund Halley (c. 1656–1743, and the discoverer of the comet that bears his name) rode out to Cambridge to try to get Newton to communicate some of his mathematical work to the Royal Society, a difficult task since Newton was very secretive and refused to publish anything. Halley said that he and his friends, the architect and natural philosopher Christopher Wren (1632–1723) and Hooke, had wondered what motion an orbiting body would traverse if attracted to a central body by a force that varied inversely as the square of the distance between them. Newton replied that in 1679 he had proven to his own satisfaction that it would be an ellipse, although he did not have the demonstration available. Halley was deeply impressed, seeing this as a breakthrough in mathematical astronomy, and urged Newton to publish. He even promised to finance the publication. Newton agreed and in an astonishingly short period produced the book that changed astronomy and physics. Published in 1687, *Philosophiae Naturalis Principia Mathematica (The Mathematical Principles of Natural Philosophy*, usually known simply as the *Principia*) laid out Newton's theory of universal gravitation and established a mathematical and mechanical model for the motion of the whole universe. In this work he used the mathematical models of Galilean physics and combined them with the planetary models of Copernicus and Kepler.

Newton had produced a universal physics, a Grand Synthesis, which finally allowed astronomers to move away from Aristotle with confidence. He had produced a real model of the universe rather than merely a mathematical set of calculations, such as Copernicus had put forward.

In the *Principia* Newton defined force for the first time, adding it to matter and motion as the third essential quality of the universe. Force, he said, was necessary to compel a change of motion. Without force, a mass (another term he originated) would continue at rest or in rectilinear motion, due to inertia. This is often called Newton's First Law. His Second Law showed the way to measure force mathematically, now expressed as $f = ma$, and his Third Law asserted that for every action there is an equal and opposite reaction. In order to understand the motion of the bodies in the universe, Newton insisted that they operate in absolute time and space. In the Newtonian system, absolute time and space were independent and unchanging aspects of reality. Absolute time was uniform, progressing at the same rate of change, everywhere in the universe and at all past and future moments. People could not perceive absolute time directly, but could infer it mathematically by observing the motion of the planets and stars. Absolute space was the unmoving and constant dimension of the universe. This meant that all observers saw the motion of the stars and planets (and everything else) within absolute space, and thus they should all see the same motion as measured against the unchanging dimension of absolute space. The implications of this were profound. It meant that Newtonian physics was truly universal and someone measuring the universe from a distant star would get the same results as we would.

Newton took the concept of centrifugal force (the tendency of an object to fly away from a circular path) and stood it on its head. In his early work in 1666, he had followed the older view of circular motion, which sought to quantify and explain this tendency to fly away. But, in a major insight, the *Principia* argued for a centripetal force instead, a force that pulls objects into the center. This is gravity. Thus, the Moon's motion is composed of two parts: first, inertial motion carries it in a straight line; second, gravity constantly forces it to fall toward the Earth. The balance of these two forces holds the Moon in orbit. (See figure 5.2.)

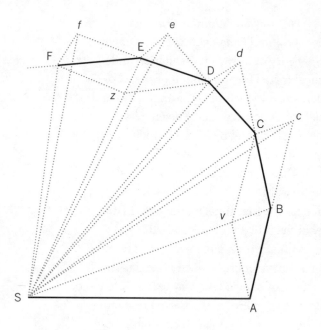

5.2 NEWTON EXPLAINS THE MOON'S MOTION

From Newton, *Principia Mathematica*, Proposition I.
At each point A, B, C, etc., the Moon's motion seeks to
move in a rectilinear line away from "S" but is drawn
back by centripetal force (gravity) toward "S."

Newton set up his argument in order to disprove Descartes's theory of vortices, which Descartes had articulated in *Principia Philosophiae* (1644). Descartes had argued that the universe was like a machine, a concept called "the Mechanical Philosophy." Except for those parts filled with coarse matter (such as the Earth), the universe was a plenum, or a space filled with an element called ether. The planets moved in a sort of whirlpool of ether that carried them along in their orbits. (See figure 5.3.)

The *Principia* was in many ways an attack on Descartes's book (the title reflected Descartes's), which is why Book Two concentrated on the study of how fluids work and how things move through them. This exploration of fluid mechanics may confuse modern readers, since it seems to have little to do with forces, gravity, or the motion of the planets. Newton's objective was not just to present his own system but to discredit Descartes's by showing that the vortex model was critically flawed.

The most far-reaching and long-lasting achievement of the *Principia* was how Newton found a way to tie together through the concept of universal gravitation the orbits of the planets and the satellites (including the Moon and comets), Galileo's law of falling bodies, the fixation of objects to the Earth, and the tides. He proved that an apple falling from a tree obeyed the same laws as the Moon orbiting the Earth. He had a law that applied to everything – the Moon, the satellites of Jupiter and Saturn, the Earth, rocks, and the distant stars.

The sheer power of Newton's universal laws suggested to scholars in many different fields that there should be similar laws governing human interaction as well. Philosophers from many other areas, including economics and political philosophy, searched throughout the eighteenth century for such laws and argued that any society that failed to follow the universal laws was doomed to failure.

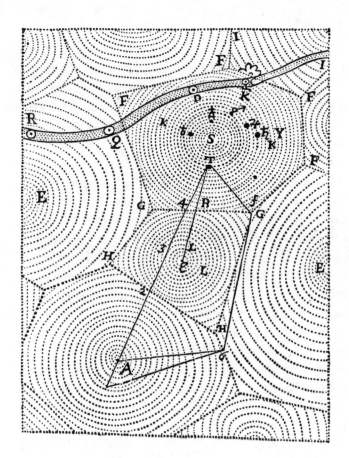

5.3 DESCARTES'S VORTEX COSMOLOGY

Points S, E, A, and ε are centers of vortices. By the churning of the corpuscles within each vortex, the center is self-illuminating, and thus is a star. Around S (the Sun) move the planets, swept along by the motion of the material of the vortex (T is terra or the Earth). The stripe at the top indicates a comet's path. From Descartes's *The World or a Treatise on Light* (1664).

Ironically, the *Principia* was not a popular or financial success. The Royal Society, the pre-eminent institutional home of English science, refused to sponsor it, since their previous publishing venture, Francis Willoughby's *History of Fishes*, had been a financial disaster. They were unable to sell more than a handful of copies of this expensive illustrated book and were forced to give it to their employees, especially Robert Hooke, in lieu of salary! The members of the Royal Society, perhaps accurately, saw the *Principia* as a book with a limited readership and so forced Halley to fund it himself, which he did through the sale of subscriptions. This story reminds us that, as revolutionary as we might now consider Newton's work, at the time it held the interest of a limited audience because so few people could understand his mathematical arguments.

Newton and Alchemy

Because Newton considered his studies esoteric, accessible only to a select few, he was not overly concerned about reaching a wide audience. One aspect of his work that has been overshadowed by his physics and mathematics and so has faded from modern consciousness was his great interest in alchemy. Both Newton and Robert Boyle were involved in alchemical investigations, looking both for the Philosopher's Stone and the basic structure and functioning of nature. Although "multipliers," or alchemists interested only in gold, were looked upon as crass, many respectable gentlemen studied alchemy. Both Newton's and Boyle's investigations came from the belief in God's ordering of the universe and were a search for the active principles that animated nature, another attack on mechanical philosophy. Since Newton was interested in how the universe worked on a microscopic as well as a macroscopic level, he investigated what the prime matter of the universe was. He read both ancient and modern authors copiously and performed experiments of long duration. Because of his expertise in this area, in later years he was appointed Master of the Mint, in which position he assessed those who claimed to have produced gold. At his death, a number of people interested in alchemy were anxious to gain access to his extensive library of alchemical works.

In spite of his lack of bestseller status and his more arcane studies, Newton was the most famous scientist of his generation. He received many important marks of honor indicative of his high status, including his position as Warden and then Master of the Mint and election as president of the Royal Society. His opulent state funeral in 1727 demonstrated his exalted position, and reports of the event were transmitted around Europe, with Voltaire (who was deeply influenced by Newton's natural philosophy) providing one of the eyewitness accounts. Voltaire was extremely impressed that a natural philosopher, and one with heterodox religious views at that, was buried with such pomp and circumstance. Here was a country that understood the importance of its intellectuals!

Mechanical Philosophy

At the same time that Newton was introducing a mathematical basis for motion throughout the universe, other natural philosophers were searching for a more concrete model. Some, looking to the new instruments and machines being developed around them, argued that the world itself was a sort of machine. As men constructed more and more precision mathematical instruments, especially for navigation, astronomy, surveying, and timekeeping, and more complex machines, this suggested to them that God might have created the most complex machine of all. This conception of the world operating in a mechanical fashion became known as the Mechanical Philosophy or, alternatively, the atomical or corpuscular philosophy. It was first developed by René Descartes and Pierre Gassendi (1592–1655). Descartes especially extended the mechanical philosophy to include living things, arguing in his *Treatise on Man* (1662) that human physiology operated just like a machine. Corpuscular philosophy was later taken up by the Englishmen Robert Boyle and Thomas Hobbes, the latter of whom used it in his political theories. Basically, it was an interpretation of the world as a machine, either as a clock (implying order) or an engine (showing the power of nature). If the universe was a machine, then God was the Great Engineer, or Clockmaker. At first, Gassendi devised this interpretation of nature to give God a role as a transcendent, rather than immanent, being. That is, Gassendi argued that God could exist outside the material universe because He had set a mechanical structure in place. There was no need for God to exist within and tinker with an imperfect world. Corpuscular philosophy was eventually accused of being atheistic because, if the universe were a perfect clock, it would never stop, and there would be no need for God. The claim that natural philosophers were removing God from the universe was unfair but widespread, affecting Newton as well as Gassendi. It recurred in many forms and as a charge against many philosophers and scientists over the next centuries.

Mechanical philosophy was rooted in ancient theories of atomism, held by Epicurus, Democritus, and Lucretius, which had been rediscovered by the humanists. These ancient thinkers had posited that the world was composed of infinitely small simple particles. For the ancient Greeks, this had shown the eternity and total materiality of the world, an aspect of the theory that Descartes and especially Gassendi (a Catholic priest) set out to change. They argued that although it appeared that the world had existed forever, this could not be true, which meant that its creation was of a different kind than its operation and was, therefore, known by God, but unknowable through natural philosophy. A material universe could not be used as a proof for or against the existence of God, since God's existence was a metaphysical, not physical, question.

Mechanical philosophers reduced matter to its simplest parts, atoms, just as Descartes had stripped away ideas through skepticism. These atoms had only two qualities: extension and motion. Since extension was a definition of matter – that is, that matter must take up space and all space must be matter – a vacuum was not possible in this philosophy. Therefore, the universe was filled with a plenum of particles. All force-at-a-distance was actually motion through the plenum, which explained magnetism and the motion of the planets. It was this aspect of Descartes's theory that Newton had attacked so forcefully in the *Principia*. And the question of whether a vacuum could exist in nature was soon taken up in Robert Boyle's experimental program. It is a great historical irony that the close examination of the nature of matter, meant to prove mechanical philosophy, refuted the very theory that the universe was full of matter.

The Use of Experiments as Proof: William Harvey and Robert Boyle

Experimentation, as a source of sure knowledge about nature, was new to the early modern period. Aristotle, of course, had argued that forcing nature into unnatural situations would tell us nothing about how it really behaved. This attitude began to change in the sixteenth century, partly because of new attitudes toward certainty and man's power over nature, and partly because skilled instrument makers were able to create precise philosophical instruments. Francis Bacon, who as Lord Chancellor had overseen the torturing of traitors, believed that human beings, when subjected to extreme pain, would be forced to tell the truth. Likewise, putting nature on trial, including torturing nature through experiments, would force her to reveal her secrets. While the reliability of truth claims as a result of experimentation was under constant scrutiny in this period, it is fair to say that one of the significant changes to the study of nature in this period of scientific revolution was the increased use of, and reliance on, experimentation.

Probably the first extended discussion of experimentation came from the study of human anatomy in the work of William Harvey (1578–1657). Following the success of Vesalius in the sixteenth century, scholars developed a keen interest in the structure and function of living things. For example, Girolamo Fabrici (c. 1533–1619) examined the structure of veins and in 1603 found the existence of valves at particular intervals. Harvey, using keen observation, some well-thought-out experiments, and a belief in the similar structure of all animals including humans (what was later called comparative anatomy), developed a theory of the circulation of the blood that proved very influential in the years that followed.

Harvey received his medical training at Padua, where Vesalius had taught, and returned to London in 1602 to work first as a physician at St. Bartholomew's

Hospital and eventually as Royal Physician to Charles I. While in these positions he conducted a series of experiments on the blood in animals. This resulted in the publication of *On the Movement of the Heart and Blood in Animals* (1628), in which he demonstrated that the blood in animals and humans was pumped out by the heart, circulated through the entire body, and returned to the heart. He proved this through a series of elegant experimental demonstrations, some involving vivisection of animals and some less invasive demonstrations with humans. In his preface Harvey drew the political parallel between this circulation and the movement of citizens around their king, indicating his close affiliation with the Royalist side of the civil war soon to erupt in England.

One of the clearest experiments that Harvey performed was his proof that there were no passages through the septum of the heart. From Greek times through Galen and Vesalius, anatomists had argued that blood must pass through the wall separating the two ventricles or large chambers of the heart. While this explanation satisfied some aspects of the assumed purpose and path of blood, problems persisted. The first was the question of volume, since the old system required the liver to produce a constant supply of blood that was completely consumed by the rest of the body. This seemed to Harvey to be out of all proportion to the amount of matter that went into the body. The second problem was purely anatomical. While Galen had said there were pores in the septum, Vesalius, who worked with human hearts rather than the cow and pig hearts that Galen had used, could find no such passages. Rather than contradict Galen completely, Vesalius argued for a permeable septum with either sponge-like tissue or pores too small to see.

Harvey reasoned that if blood were circulated, a far smaller volume would be needed; it was simpler to conceive of a reused supply of blood than a constantly created and consumed supply. At the center of the system was the heart, working as a pump, but to demonstrate the two-part circulation (heart-to-lungs and back, heart-to-body and back) he had to show that blood did not move from the venous to the arterial system by way of the chambers of the heart. In a later experiment he demonstrated this, using a cow heart and a bladder of water. He tied off the passages to the heart so that he could squeeze water into one ventricle and see if it passed into the other. When it did not, he had the first proof that the blood did not pass through the septum. He then tied or untied the constraints and used the bladder of water to show that the blood must pass in sequence from right atrium to right ventricle and out to the lungs, and then through the pulmonary vein to the left atrium and left ventricle and out to the body through the aorta. While it

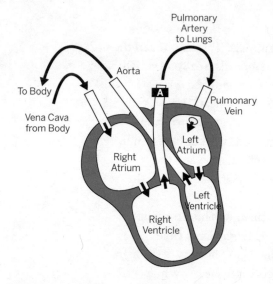

Pulmonary
Artery
to Lungs

Aorta

To Body

Vena Cava
from Body

Pulmonary
Vein

A

Right
Atrium

Left
Atrium

Right
Ventricle

Left
Ventricle

5.4 HARVEY'S MODEL OF THE HEART

Harvey closed off the artery at "A" and
pumped water through the vena cava to
demonstrate that no fluid passed from
the right ventricle to the left ventricle. This
disproved Galen's model of the heart.

was still unclear how the blood got from the
arteries through the tissues of the body and back
into the veins (and would remain unclear until
microscopy developed far enough to see the
microvessels at cell level), Harvey's work better
explained the evidence than the older system.
(See figure 5.4.)

Harvey also used careful observation and some
experimentation in his embryological studies.
Fabrici, in his book *On the Formation of the Egg and
the Chick* (1621), had observed that in viviparous
generation the embryo was created by a union of
semen and blood from the parents. Harvey followed
this work with a close examination of the development of ova. He examined fertilized eggs from their
unformed state to birth, tracing even more closely
the stages of growth. He published this work in
Exercises Concerning the Generation of Animals (1657).
This spurred further work by Marcello Malpighi
(1628–94), who introduced the use of microscopic
observations of ova development in 1672.

The work of Harvey and Malpighi, showing the power of observation and
experiments, accorded with the inclination among many natural philosophers to
use instruments and demonstrations to isolate phenomena and break down
investigations into smaller components. Perhaps the man most responsible for the
elevation of instrumental investigation during the scientific revolution was Robert
Boyle (1627–91). Boyle, the son of an Irish noble family, came to Cambridge
during the English Civil War and became a key player in the creation of the Royal
Society after it. He joined the elite of London society from the 1660s onwards,
living with his sister, Lady Ranelagh, in her house in Pall Mall, and entertaining
royal and scholarly guests alike at his laboratory there. While at Cambridge, he
set up his own alchemical laboratory to investigate the underlying make-up of
matter. He denounced old-fashioned alchemical investigations and, in *The
Skeptical Chymist* (1661), laid a foundation for the new study of chemistry.

Boyle, with Robert Hooke as his assistant, investigated airs, using a newly
devised air pump. He employed instrument makers to attach a large, carefully
blown glass globe to a pump in order to evacuate the air from the globe. In this,

5.5 ILLUSTRATION OF VON GUERICKE'S MAGDEBURG EXPERIMENT

he followed the lead of Otto von Guericke (1602–86), who had performed his own air experiments in the 1640s. Most famously, in 1657, von Guericke demonstrated that two teams of horses hitched to joined hemispheres of copper (the so-called Magdeburg hemispheres) from which the air had been evacuated could not pull them apart, since the weight of the air outside the spheres was so much greater than that on the inside. (See figure 5.5.)

Boyle and Hooke built their first air pump in 1658 and performed a number of experiments. (See figure 5.6.). They demonstrated that air had weight, that a vacuum could exist, and that some component of air was necessary for respiration and combustion. Their results were published in 1660 as *New Experiments Physico-Mechanical Touching the Spring of the Air and Its Effects*. Boyle used his air pump for less spectacular demonstrations than von Guericke, for example, placing small animals in the glass sphere, removing the air, and watching them perish, or alternately placing candles therein and watching the flame extinguish. (See figure 5.7 for a later depiction of these events.) From these experiments he concluded that there was something in the air that supported both life and combustion. His work was thus connected to Harvey's because it touched on what part of air was needed for life and that it seemed to be brought into the body by way of the lungs. This was part of a growing interest in "vitalism," the search for the spark of life that transformed inanimate matter into living plants and animals.

5.6 BOYLE'S AIR PUMP AND TOOLS FROM *NEW EXPERIMENTS PHYSICO-MECHANICALL* (1660)

Boyle's work also demonstrated the relationship between pressure (the "spring" of the title) and volume of air. He and Hooke used a j-shaped tube filled with mercury to show that increasing or decreasing the pressure on the short stem raised or lowered the level of mercury in the long stem. (See figure 5.6.) Boyle argued that "according to the *Hypothesis*, that supposes the pressures and expansions to be in reciprocal proportion";[1] in other words, as pressure goes up, the volume of air goes down in equal proportion and vice versa. While Boyle was making a specific argument about atmospheric air, which he considered an elastic fluid, not an element in itself, the basic relationship he pointed out was later transformed into $PV = K$, what we now call "Boyle's Law" or occasionally "Mariotte's Law" after Edmé Mariotte (1620–84), who independently found the same relationship in 1676.

Unfortunately, Boyle's air pump was plagued with problems. It leaked quite badly, so that it was not really possible to evacuate all the air from the interior. Also, although he published detailed accounts of his instrument and its operation, complete with schematic diagrams, other natural philosophers across Europe could not replicate his results. Thomas Hobbes (1588–1679), a natural philosopher as well as a moralist, severely criticized the work of Boyle and Hooke. Hobbes claimed that the air pump did not work and that it in no way represented a vacuum, as they had claimed. Although Hobbes presented many sound arguments, Boyle's rising status, both socially and scientifically, ensured that his was the winning side of this disagreement.

....

1.   Robert Boyle, *A Defence of the Doctrine Touching the Spring and Weight of the Air* (London: M. Flescher, 1682), 58.

5.7 *AN EXPERIMENT ON A BIRD IN AN AIR PUMP*, JOSEPH WRIGHT OF DERBY (1768)

Source: National Gallery, London, UK / Bridgeman Images.

Despite equipment problems, the claim of replicability became fundamental to the experimental program from Boyle on. That is, natural philosophers began to claim that experiments were useful and produced true and certain results precisely because they were not dependent on the experimenter but could be repeated by anyone. The air pump was to be seen as transparent; nothing stood between the observer and the aspect of nature being observed. In other words, an ideology of the objectivity of instrumental experiment arose from Boyle's work with the air pump. Our modern reliance on experimentation – and on replicability – is in part based on the triumph of Boyle and the Royal Society over Hobbes and other skeptics. That "proof" (that is, evidence of the true state of nature) results from experimentation was, and is, a powerful idea, even if it has philosophical flaws. Many things make an unmediated view of nature impossible, but because of Boyle's status and the status of his witnesses, the air pump became "black-boxed." That is, the air pump and other experimental instruments were considered neutral and objective, unproblematically revealing nature's state. For example, when we look at a thermometer to help us decide what kind of coat to wear outside, we have accepted a particular concept of temperature. The thermometer is not an unbiased object but embodies a philosophical idea about the quantification of nature. To someone unfamiliar with the concept of temperature, a thermometer would be a meaningless device. This does not mean that what the thermometer reveals is false but that all scientific instruments represent a system of beliefs about the world.

The Development of Philosophical Instruments

During the seventeenth century, other instruments (often called philosophical instruments) and experimental programs were devised, all sharing this ideological stand with Boyle's work. Boyle himself used the barometer, developed in 1644 by Evangelista Torricelli (1608–47), and others followed this lead in investigating the weight of air. Torricelli filled different glass tubes with quicksilver (mercury), inverted them into a basin, and discovered that all the tubes maintained a constant level. Torricelli claimed that the space above the mercury, the "Torricellian space" as Boyle called it, was a vacuum and argued that "we live submerged at the bottom of an ocean of the element air, which by unquestioned experiments is known to have weight."[2] He also predicted that if one ascended to higher altitudes, the weight of air would be less and the column of mercury would descend further. This prediction was taken up by the mathematician Blaise Pascal. Pascal first worked to replicate Torricelli's instrumental experiment, which proved a difficult task. He eventually succeeded, both with a column of mercury and with a much larger one of water. This resulted in a heated debate about the possibility of a vacuum, strongly denied by a number of leading theologians. To avoid this discussion and concentrate instead on the question of the weight of air, in 1648 Pascal took the barometer up a mountain near his brother-in-law's home in Clermont, France. Sure enough, the higher the ascent, the lower the column of mercury and the larger the "Torricellian space" at the top of the

----

2. Evangelista Torricelli, "Letter to Michelangelo Ricci, 1644" in *Encyclopedia of the Scientific Revolution*, ed. Wilber Applebaum (New York: Garland, 2000), 647.

tube. Because Pascal soon thereafter had a crisis of
faith and turned from natural philosophy to spiritual-
ity, the results of this investigation were not known
until after his death, with the posthumously published *Traités de l'équilibre des liqueurs et de la
pesanteur de la masse de l'air* (1663). This work, with
Boyle's, set up an experimental research program for
the coming century, as well as demonstrating the
power and "objectivity" of philosophical instruments.

Perhaps the most innovative philosophical
instrument of the seventeenth century was the
microscope, first developed in the first decade of the
century. Following the success of the telescope to
bring distant sights closer, unknown instrument
makers, probably in the Netherlands, produced
instruments designed to greatly magnify the very
small. The five microscopists best known for their
startling observations and discoveries were van
Leeuwenhoek, Hooke, Malpighi, Swammerdam, and
Grew. Antoni van Leeuwenhoek (1632–1723), a
merchant in Delft, first turned these magnifying
devices on a variety of substances, most famously
male semen, where he claimed to observe small

5.8 ILLUSTRATION FROM ROBERT HOOKE'S
MICROGRAPHIA (1665)

animalcules in motion. This resulted in a series of interesting letters to the Royal
Society. Robert Hooke turned to the favorite subjects of microscopists – insects,
seeds, and plants – and captured some stunning images of various enlarged
phenomena through engravings in his bestselling, lavishly illustrated book,
Micrographia (1665). (See figure 5.8.)

The Italian anatomist Malpighi turned the microscope on the human body and, in
addition to his embryological work, discovered capillaries and their role in the circulation of the blood. Jan Swammerdam of Amsterdam (1637–80) disproved contemporary
theories about the metamorphosis of insects, while the Englishman Nehemiah Grew
(1641–1712) found the cellular structure of plants. All five successfully overcame early
suspicions that the instrument was creating and disguising as much as it was revealing
to produce theories and observations much sought after by the natural philosophical
community. By the eighteenth century the vanishingly small was seen to be observable,
without scientists worrying about any interference from the apparatus itself.

Newton and the Experimental Method

Isaac Newton also played a part in the development of experiment (and experimental instruments) as a legitimate methodology. Beginning during his *annus mirabilis* he developed a theory of light based on a series of simple and elegant experiments. Because of a decades-long dispute with Hooke about optics, he refused to publish, but in 1703 Hooke died. Newton's *Opticks* came out in 1704. Unlike the *Principia*, the *Opticks* was written in English, in simple language, and laid out so that the experiments could be recreated by anyone who could read the book and afford a few pieces of optical equipment such as prisms, mirrors, and lenses. Even more than Harvey or Boyle, who had likewise explained their procedures in print, Newton became the model for the new experimental method. It was a smash hit, snapped up by an eager public and translated into French, German, and Italian within the year. It has been in print almost continuously to the present day.

Newton was contributing to a long tradition of optics research, stretching back to the Middle Ages and earlier Arabic natural philosophers. He was also building on the work of Kepler, who had argued that light traveled in rectilinear rays, enabling a mathematical description of its path. This allowed several natural philosophers, including Thomas Harriot (c. 1560–1621), René Descartes, and Willebrord Snell (1580–1626), to develop the sine law of refraction, which stated that when a ray of light passes from one transparent medium to another (such as air to water), the sine of the angle of the incident (original) ray divided by the sine of the angle of the refracted ray equals a constant. While this was a useful and demonstrable relationship, it led to a disagreement about the nature of light. Was it a motion through matter (that is, a wave)? Or was light made of particles? How did light travel? Could it travel in a vacuum? During the 1670s Huygens developed a wave theory of light in which he argued that the idea of a wave front represented the path of the light.

Newton criticized this wave theory, largely because it seemed to contradict the rectilinear nature of rays put forward by Kepler. Newton argued that light was corpuscular and proved it to his own satisfaction with a series of elegant experiments. By passing a beam of sunlight through a series of prisms, he demonstrated that white light was not pure light, as had been supposed, but rather was a composite of many colors (the spectrum). His demonstration has been called the *experimentum crusis* or crucial experiment, the demonstration that confirms the hypothesis. What Newton noticed was that light passing through a prism smeared

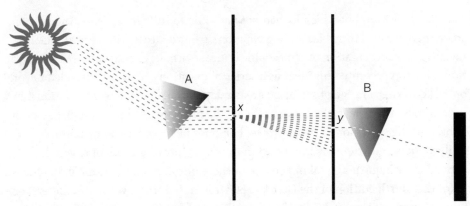

5.9 NEWTON'S DOUBLE PRISM EXPERIMENT

Sunlight passes through the prism "A" and falls on the screen. A portion of the spectrum produces light of a single color that passes through the screen at slit "x" and continues on to the second screen, which goes through slit "y" and on to the second prism "B." The light is finally projected on to the wall. Newton's experiment demonstrated that light of a single color could not be further broken up into the spectrum and thus white light was actually a mixture of colors.

into an oblong with colors at top and bottom. It had been assumed since antiquity that such an effect was the result of some corruption of the white light. If this was the case, it seemed reasonable to assume that by passing some colored light through a second prism, the degree of corruption would be increased. So, Newton passed light through a prism, then allowed a small amount of colored light to project through a slit, and then through a second prism. There was no change in the color of the light. In other words, nothing was added to or taken away (see figure 5.9). To confirm this, he also placed two prisms together so that the first separated the light and the second, an inverted prism, gathered all the bands back together, producing a spot of white light.

Newton believed that light was composed of particles and that their differing speeds resulted in a differing angle of refraction when passed through a prism. If, for example, all red light was composed of small particles of similar nature, there seemed to be no mechanism to change the nature of the red particles as they passed through successive prisms.

This was hotly debated, the French following Descartes's and Huygen's lead in preferring a wave theory of light, but Newton's *Opticks* provided a foundation for a new English school of optical research throughout the eighteenth century.

Newton's *Opticks* also laid out a research program for natural philosophers who followed him. The book concluded with a series of "Queries," topics that interested Newton but which he had not had time to fully investigate. In addition, since there was a strong bias among Newton's peers against "theoretical" science, meaning the presentation of philosophical ideas without experimental demonstration, Newton presented his ideas as a series of questions. This list extended beyond optics, covering a range of scientific areas related to light such as the relationship between light and heat, the effect of the media of transmission on the behavior of light, and the condition of the universe. For close to 100 years many natural philosophers and scientists interested in finding important areas of research started their inquiries by taking on one of these questions. For example, in Query 18 (of the fourth edition of the *Opticks*), Newton noted that two thermometers, one in a vacuum and the other in a closed container of air, both seemed to heat up and cool down at about the same rate. This suggested to Newton that there must be a medium of propagation "more rare and subtle than the Air" that transmitted heat like a vibration. This observation prompted a number of scientists to look for the imponderable fluid or ether, an idea that was eventually resolved by the work of Einstein. It also got other scientists to investigate the nature of heat as separate from temperature, leading ultimately to the kinetic theory of heat and the laws of thermodynamics.

One of the most famous questions posed by Newton was Query 31 from the 1718 edition. Newton asked: "Have not the small Particles of Bodies certain Powers, Virtues, or Forces by which they act at a distance, not only upon the Rays of Light for reflecting, refracting, and inflecting them, but also upon one another for producing a great Part of the Phenomena of Nature?" Newton goes on to suggest that known forces such as gravity and electricity may play a role in the attraction of particles, but unknown forces may also be at work. By observing attraction, according to Newton, it should be possible to figure out the law of attraction that controls the way matter combines and functions. This idea contributed to the concept of chemical affinity; an idea used by most chemists to explain how matter combined. Affinity theory was one of the foundational concepts of modern chemistry until the nineteenth century when it was replaced by valence theory.

New Scientific Organizations

The new instruments, experiments, and the underlying assumptions about the relationship between humans and nature were fundamentally important to the creation of a new scientific enterprise in the seventeenth century. Equally important were the new institutional structures that developed in this period for the express purpose of fostering and supporting natural philosophy. The founding of such assemblies as the Royal Society of London and the Académie Royale des Sciences[3] in Paris contributed to a dramatic new organization of science, one that encouraged natural philosophers to develop social codes of behavior, rules about who could do science and what counted as science, and statements about the secular usefulness of their enterprise. Just as Galileo had tried to separate religious and scientific claims to the monopoly of truth, so too did these new scientific societies. This was the beginning of the institutionalization of science, different in kind from the university-based science and distinct from the court-based science of the sixteenth century, although owing much to that earlier model. The breakdown of individual princely courts cast some natural philosophers adrift. Growing absolutism, especially in France, led to a more particular focus for patronage, as well as to a growing urban elite and intellectual culture in the capital cities. One hundred and fifty years of bloody religious wars caused people to look elsewhere than the church for salvational knowledge and for a secure group with status. The rising leisured class was looking for secular legitimation and something to do.

In about 1603 the first secular scientific society was formed, the Accademia dei Lincei (Academy of the Lynx) in Rome. This organization was founded by Federico Cesi (1585–1630), later Prince Cesi, and existed in great secrecy for the first years of its life, persecuted by the authorities including Cesi's father. Cesi was a man of huge intellectual energy and curiosity, who established a research program in natural history of great scope and imagination. When Galileo joined in 1611, the Accademia was given a major boost, especially since Galileo donated his microscope to it. After the condemnation of Copernicus, however, followed by Cesi's death in 1630 and Galileo's condemnation, the Accademia ceased to function. In 1657 a scientific society was founded in Florence, the Accademia del Cimento (of experiments). This society had neither a formal membership nor statutes and existed for ten years as a loose collection of men interested in experimental research. It, like the Lincei, was in reality a hybrid, neither court-based nor autonomous, since it was focused on the court of Grand Duke Ferdinand II de Medici and his brother, Prince Leopold.

....

3. The name of this organization changes according to informal or official use. Historians say "Académie des Sciences," although officially it is called Académie Royale des Sciences. To make things more complicated, the name changed several times even though it remained the same group. Within the text, we will use the term commonly used by historians, that is, the Académie des Sciences.

The most famous of the seventeenth-century scientific societies, the one that first marked this new form of scientific organization and the only one that has survived as a continuously operating body to the present day, was the Royal Society of London, established by royal charter in 1662. There is much debate about its origins. A number of people interested in natural philosophy, experimentalism, and the utility of natural knowledge met informally in England during the English Civil War and in the Interregnum that followed. The Civil War was fought between Royalists, who wished to maintain the power of the Crown and the liberal Anglicanism of the Church of England, and Parliamentarians, who argued for the primacy of political power from the people in Parliament and for the predestinarian theology of the Puritans. Historians of the Royal Society have sought founders of modern science in both camps, but most particularly among the Puritans. Samuel Hartlib (c. 1600–62), an educational reformer who came to England from Prussia to escape the Thirty Years' War and who was closely associated with the Parliamentary camp, tried to develop an educational program to bring natural philosophy and the new experimental method to English intelligentsia. The Hartlib circle was one of the groups responsible for the founding of the Royal Society after the Restoration, although on much more conservative and elite lines than Hartlib and his circle had envisaged. In fact, while earlier historians sought the origins of this new institutional structure in radical religion and politics, the truth seems to be that most natural philosophers were eager to find an alternative to the crippling political and religious controversies of their day and that natural philosophy provided just such a third way.

With the Restoration of Charles II in 1660, various groups from

5.10 FRONTISPIECE FROM SPRAT'S *HISTORY OF THE ROYAL SOCIETY* (1667)

London, Oxford, and Cambridge came together in London to form the Royal Society. Although it received a charter from the king, it was autonomous and, therefore, different from earlier university, church, or court-based spaces of scientific investigation and discussion. It was founded with a strong inclination toward a Baconian philosophy of research, setting out to employ an inductive, cooperative method in order to discover useful information for the benefit of the nation. Here the rhetoric of utility that emerged with the courtly philosophers was carried into an urban, gentlemanly locale. Thomas Sprat, the first official historian of the Royal Society, said that the Society was founded as a way to avoid the "enthusiasm" of the Puritans and the sectarian disputes that had ripped the country apart during the Civil War. Although Sprat was hardly a disinterested observer, it does seem that the Royal Society attempted to find a third way through the religious and civil disagreements of the period. While Royal Society members included those of many different religious affiliations (from Catholic to Puritan), what they had in common was the desire to keep clear of religious controversy and to do natural philosophy instead of theology.

The Royal Society developed a strict method of choosing members, who had to be known to existing members and have an active interest in natural philosophy. The exception to this were some aristocrats, necessary to maintain the Society's elite nature. It also refused to allow women to join, although Margaret Cavendish, Duchess of Newcastle (1623–73), did attend some meetings and had published more books on natural philosophy than many of the members put together. They were also very hesitant to accept tradespeople, preferring instead the trustworthiness of gentry. The Royal Society developed a gate-keeping function, determining who counted in natural philosophical inquiry. Through the work of their first corresponding secretary, Henry Oldenburg (c. 1619–77) and through the publication of their journal, the *Philosophical Transactions of the Royal Society*, founded in 1665 and still published today, they also were able to determine what counted as proper natural philosophical work. Thus, in one fell swoop, the Royal Society became the arbiter of just who could be a natural philosopher and what qualified as acceptable natural philosophy.

The other successful seventeenth-century scientific society was established in a very different way. The Académie Royale des Sciences was founded in Paris in 1666 by Louis XIV's chief minister, Jean-Baptiste Colbert. Although there had been an informal network of correspondence centered on Father Marin Mersenne (1588–1648) since the 1630s, the Académie was a top-down organization, another element of the absolutist French state. Unlike the Royal Society, where members were elected and unpaid, the state appointed 16 academicians, paid as civil servants, to investigate the natural world as the king and his advisors required. So the Académie can be seen as the root of the professionalization of science, since this was the first instance where scholars were paid exclusively as scientists. Because their research agenda was set by the state, they could take on projects beyond the scope of individual scientists. For example, the Académie sponsored the measurement of one minute of arc of the Earth's surface, resulting in the first accurate measurement of the size of the Earth and the distance to the stars. In the long run, however, it was less successful than the Royal Society. An appointment as an Academician was often the reward for a life's work rather than an incentive to new work. Most of the massive projects came to nothing. Its journal, *Journal des Sçavans* (founded in 1665), was largely a reprint service. The Académie des Sciences did well as a promoter of science as an elite and respected activity, but it was not a place that sponsored innovation.

These new scientific societies created four enduring legacies for science as a profession. First, science was now seen as a public endeavor, although with carefully defined limits, members, and methods. Second, its cooperative nature was stressed, through projects such as the History of the Trades sponsored by the Royal Society and the History of Plants and Animals investigated by the Académie. Such undertakings led to the Enlightenment view that all was knowable if properly organized and to the sense of the utility of the knowledge gained. When Leibniz founded the Berlin Academy of Science in 1700, he chose as its motto *theoria cum praxi*, theory with practice.

Third, scientific communications were established as an essential element of the scientific enterprise. While it had been true that letters between natural philosophers (for example, between Galileo and Kepler) or within the letter-writing circle of Father Mersenne had been integral to maintaining a community of scholars, the establishment of scientific journals in the seventeenth century both broadened and controlled this community. These journals acted as a guarantor of veracity and reliability, even while issues were socially determined and highly contested. They also broadcast scientific ideas and experiments to a much larger audience, allowing ordinary people to take part in science by "virtual witnessing." This led to a wider interest in the investigation of nature and a greater acceptance of new scientific ideas and of scientists as respectable, if awe-inspiring, people.

Finally, scientific societies established scientists as experts, qualified by membership to pose and judge questions about nature. This was especially true in France, where election to the Académie was the culmination of one's life's work. But equally within the Royal Society, some natural philosophers, such as Newton or Boyle, were respected both within and outside the society as experts and significant scholars. Below them were the collectors, those who found interesting natural phenomena to report but who left theorizing to their betters, much as Bacon had laid out in Solomon's House. Thus, the seventeenth-century scientific societies established ideologies about science and its practice is still with us today.

During the sixteenth century, connections between natural philosophers and skilled artisans aided greatly in the development of new ideas about nature and new problems to be investigated. This was also true during the seventeenth century, although the focus became the urban mercantile center, rather than the princely court. Just as the scientists themselves were forming associations with their fellow scholars in these new, secular, non-court settings, often in major trading centers, they were also closer to shipyards, print shops, instrument makers, and chart makers. However, projects like the Royal Society's History of the Trades, an unmitigated disaster, shows us that the communication between these artisans and scholars was more complicated than one might think. The History of the Trades attempted to find out how all the different manufacturing trades in England were performed, so that natural philosophers might find a more rational scientific way to manufacture goods. Not surprisingly, the tradesmen were remarkably unforthcoming about their trade secrets, and the suggestions made by the bewigged gentlemen were at best unhelpful and at worst positively dangerous. It would take some time for a new collaborative approach to bring together the skills of craftspeople and the precision of scholars.

The Place of Women in the Study of Science

Most of this discussion of seventeenth-century science has focused on the contributions of men of science; this was, however, a period when women were attempting to make their mark on the study of the natural world. Both Anne Finch, Viscountess of Conway (1631–79), and Margaret Cavendish, Duchess of Newcastle, were interested in breaking into this previously clerical and male preserve. The same social and intellectual upheaval that made science a gentlemanly pursuit gave women a brief window of opportunity to become involved in natural philosophy. Bethsua Makin (c. 1612–c. 1674) wrote *An Essay to Revive the Ancient Education of Gentlewomen* (1673), arguing for the right and ability of women to study natural philosophy. Anne Conway corresponded with Leibniz and shared with him her theory of "monads" that became the basis of his particulate philosophy of the universe. Noblewomen such as Christina of Sweden engaged in natural philosophical conversations. Margaret Newcastle wrote many books of natural philosophy and attended a meeting of the Royal Society. She also composed what has been called the first English work of science fiction, *The Description of a New World, Called the Blazing World* (1666). These women, however, were exceptions. The seventeenth and eighteenth centuries saw restrictions on women's sphere in science as in much else.

Such restriction came from a general change in attitude to a gendered nature and from changing theories about women's role in society and in reproduction. In a pre-industrial society the vast majority of people existed in a close and symbiotic relationship with nature, which was perceived as female, a nurturing mother. The ideal was coexistence, not control. Mining, for example, was either the rape of the Earth or the delivery of a child, since minerals developed in the Earth's womb. Therefore, sacrifices, prayers, and apologies were necessary before mining could begin. The vitalism of Paracelsus, the natural magic of the neo-Platonists, and the naturalism of Aristotle all gave the female contribution of nature and to nature its due.

In early modern Europe, however, this began to change. As scholars began to view the Earth as exploitable, its image changed to a wild female who must be tamed. At the same time women were losing socioeconomic status, becoming less autonomous and less able to earn wages or operate independently in craft guilds. Increasingly, accusations of witchcraft were leveled, especially against women. Scientific theories of sexuality and procreation changed. Where during the Middle Ages women had been recognized as providing something to initiate procreation (matter for Aristotle, female semen for medieval authors), during the sixteenth and seventeenth centuries theorists such as Harvey claimed that women were merely a receptacle, the incubator of offspring, and, as such, were totally passive. Likewise, women were no longer seen as equal partners in intercourse but rather seducers, enticing men into sexual activity that was deleterious to their health and intellectual well-being. Women were also losing their professional role in reproduction, as female midwives were replaced by licensed male surgeons or male midwives, who used their new technology of forceps to manage nature. Finally, with the introduction of the Mechanical Philosophy, introduced partly to deal with perceived disorder in the world, the soul of nature was stripped away, leaving only inanimate atoms; nature was dead, and women's claim to vitality through reproduction was rendered void. In other words, the natural philosophy of the seventeenth century articulated an ideology of exploitation, an image of the world that could be constructed according to man's specifications.

We can trace this change in attitude toward women as natural philosophers through the careers of two scientists: Maria Sybilla Merian (1647–1717) and Maria Winkelmann (1670–1720). Their careers demonstrate, on the one hand, the possibility of women's involvement in natural inquiry and, on the other, the restrictions to their participation through the creation of the new institutions of scientific societies.

Maria Merian's career illustrates the success a woman could have in a scientific field, particularly one based on an entrepreneurial model. Merian was born in Germany into a family of artists and engravers. From an early age she was interested in drawing and painting insects and plants. After marrying her stepfather's apprentice she became a renowned insect and plant illustrator, publishing well-received and beautifully engraved books. In 1699 the city of Amsterdam sponsored her travels to Suriname, where she observed and recorded many new plants and animals. On her return she created a bestselling book of these findings, which was published posthumously. Merian's career thus followed a very successful path in the older apprenticeship and mercantile model. However, after her death her work came to the attention of the new community of natural historians and philosophers, and her reputation suffered. Her Suriname book was strongly critiqued, condemned for her classification system and more particularly for her credence of slave knowledge about the use of these plants and insects. Her reputation, therefore, was greatly diminished in the increasingly misogynistic and racist attitudes of the eighteenth-century scientific community.

Maria Winkelmann was the daughter of and later wife of astronomers (she married Gottfried Kirch in 1692) and worked with both her father and husband on telescopic observations in Berlin. In 1702 Winkelmann independently discovered a comet and published her findings. She was a full participant in her chosen scientific field, but after Kirch's death in 1710 her status fell sharply. The Royal Academy of Sciences in Berlin refused to allow her to continue in her husband's position as official astronomer to the Academy, eventually appointing her less capable son instead. Even Leibniz's support was insufficient to help her maintain her position; the Academy was unwilling to set a precedent by allowing a woman to hold such an important job. In this they followed the lead of the Royal Society, which had likewise debated allowing Margaret Cavendish to join and had resisted the undesirable precedent. Eventually the Academy, embarrassed by her presence at the observatory, forced her to leave the premises; without access to large telescopes, she was unable to continue her observational work. The new organization of science, the Academy, proved itself to be more restrictive for women than the earlier apprenticeship model had been.

Seventeenth-Century Scientific Ideology

The new ideology of exploitation of and superiority toward nature reflected a changing attitude toward knowledge and nature. As crucial as the new knowledge and methodologies proved to be, of even greater significance in the creation of modern science were the new locales of scientific discussion and the new ideology and code of conduct the seventeenth-century societies established. Science had previously been the property of clerics and academics, but the upheaval of the seventeenth century – its religious and political wars, its economic strife – created an opportunity for a new group of gentlemen practitioners to develop a new standard for scientific conduct and a new place to practice science. This was particularly true in England, where the move away from absolutism at the end of the century allowed a certain freedom of association among the leisured classes and where the controversies associated with the Civil War and its aftermath gave gentlemen a desire for civility and an alternate way to establish matters of fact.

Robert Boyle was particularly instrumental in this transformation of scientific ideology. First, he set up laboratories, spaces for scientific experiment and investigation, in private locations, particularly in his sister's townhouse in fashionable Pall Mall in London. The privacy of this space was paramount, since Boyle was able to control access and behavior within the site. That is, he could allow in people with the proper credentials, worthy to witness his experiments and guaranteed to know the proper way to behave. Similarly, private museums of natural history developed in this period all over Europe as private spaces, also belonging to aristocrats and gentry, who controlled visitors and bestowed on those visitors status through their permission to observe.

Because this space was private, Boyle could decide who had the proper credentials to observe, to take part in experiments, and to participate in the making of natural knowledge. He developed a number of criteria, later used by the Royal Society and other scientific bodies. The person seeking entry had to be known to Boyle or his circle and, thus, was a gentleman. This person should also be a knowledgeable observer, one who was able to validate the experimental knowledge and to witness matters of fact, rather than just gawk. Still, it was more important that the observer be of the leisured classes than that he or she be philosophically knowledgeable. This was clear from von Guericke's famous hemisphere experiment, which gained its status as one that created natural knowledge about the weight of the air by the presence of a large group of gentry onlookers (as can be seen in figure 5.5). Similarly, one of the important roles of the Royal Society was to provide a credible audience for various experimental demonstrations, thereby establishing their veracity.

Because these experiments and demonstrations of knowledge were performed in spaces created and used by gentlemen, gentlemanly codes of behavior were adopted as the codes of behavior for scientists. For example, gentlemen argued for the openness and accessibility of private space at the same time that they were carefully controlling access to such places. Likewise, scientific laboratories as they developed claimed to be open public space while limiting access to those with the knowledge and credentials to be there. Gentlemen were very concerned with issues of honor and argued that their word was their bond. A gentleman never lied, which was why there was much outrage (and duels fought) at any suggestion of cheating. This was why matters of fact could be established through the witnessing of a few gentlemen, who would, of course, see the truth of the investigation and report it accurately to others. A scientist's word was also his bond. Scientists would not cheat or lie, and thus they claimed the role of a completely trustworthy enclave in society. The Royal Society in concert with Boyle created a community of scientists who could decide what constituted truth and reality and who were allowed to make pronouncements on that reality.

Conclusion

SIR ISAAC NEWTON.

5.11 NEWTON, BASED ON THE LOUIS FRANÇOIS ROUBILIAC BUST OF NEWTON AT TRINITY COLLEGE, CAMBRIDGE

By the time Newton died in 1727, the place of the natural philosopher had changed considerably. The image of the natural philosopher had shifted even further away from the "pure" intellectuals of ancient Greece, the Islamic wise men, or even the courtiers of Galileo and Kepler's era. A bust commemorating Newton sits in the entrance hall of Trinity College, Cambridge. (See figure 5.11.) It does not present him as an erudite scholar or as an important member of society, as had earlier portraits from his student days, or as Master of the Mint. Rather, it presents him as a modern Caesar who, with firm gaze and noble brow, has conquered all he surveyed. While Caesar captured Rome and gained an empire, Newton conquered Nature and made it man's dominion. The poet Alexander Pope said of Newton's life: "Nature and Nature's Laws lay hid in night / God said, 'Let Newton be,' and all was light."[4]

By the end of the seventeenth century many aspects of modern science had been established. Philosophers of this period of scientific revolution had wrestled with questions of epistemology at the beginning of the century and decided on a behavioral model of truth-telling by the end. Some of these thinkers developed new ideas, theories, and experimental discoveries, setting in place a series of research programs for the coming century. They had also introduced a new methodology of science, which included the mathematization of nature and a new confidence in, and reliance on, experimentation. New secular scientific institutions sprang up; with them came an articulation of the utility of their knowledge to the state and the economy. Finally, the domination of natural philosophical inquiry by secular gentlemen from the leisured classes ensured a code of behavior that was gendered and class-based. These ingredients led to a new scientific culture that rapidly assumed a recognizably modern face. All these changes together constituted a scientific revolution.

....

4. Alexander Pope, "Epitaph Intended for Sir Isaac Newton," in Westminster Abbey, 1735.

Essay Questions

1. Was there a scientific revolution? If so, of what did it consist?

2. In what way can Newton be seen to have completed the Copernican revolution in astronomy?

3. In what ways were the Royal Society and the Académie des Sciences similar and different?

4. What role did women play in the development of natural philosophy and did this change over time? Why or why not?

CHAPTER TIMELINE

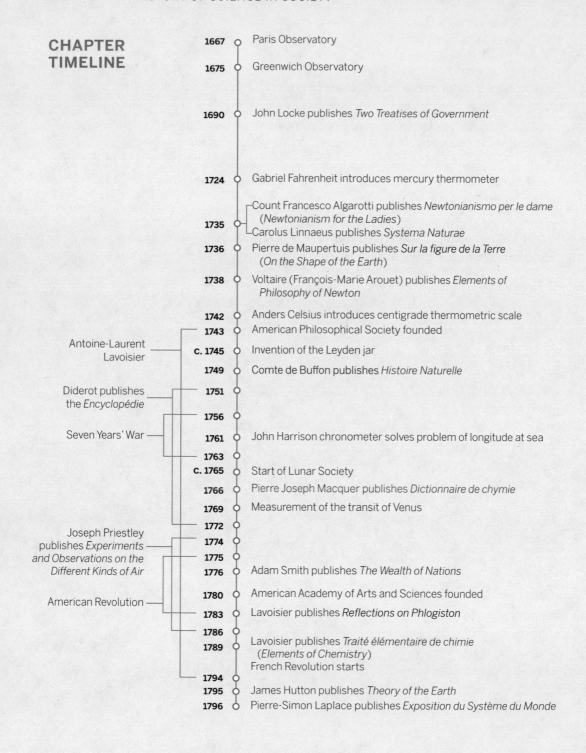

1667	Paris Observatory
1675	Greenwich Observatory
1690	John Locke publishes *Two Treatises of Government*
1724	Gabriel Fahrenheit introduces mercury thermometer
1735	Count Francesco Algarotti publishes *Newtonianismo per le dame* (*Newtonianism for the Ladies*)
	Carolus Linnaeus publishes *Systema Naturae*
1736	Pierre de Maupertuis publishes *Sur la figure de la Terre* (*On the Shape of the Earth*)
1738	Voltaire (François-Marie Arouet) publishes *Elements of Philosophy of Newton*
1742	Anders Celsius introduces centigrade thermometric scale
1743	American Philosophical Society founded
c. 1745	Invention of the Leyden jar
1749	Comte de Buffon publishes *Histoire Naturelle*
1751	
1756	
1761	John Harrison chronometer solves problem of longitude at sea
1763	
c. 1765	Start of Lunar Society
1766	Pierre Joseph Macquer publishes *Dictionnaire de chymie*
1769	Measurement of the transit of Venus
1772	
1774	
1775	
1776	Adam Smith publishes *The Wealth of Nations*
1780	American Academy of Arts and Sciences founded
1783	Lavoisier publishes *Reflections on Phlogiston*
1786	
1789	Lavoisier publishes *Traité élémentaire de chimie* (*Elements of Chemistry*)
	French Revolution starts
1794	
1795	James Hutton publishes *Theory of the Earth*
1796	Pierre-Simon Laplace publishes *Exposition du Système du Monde*

Antoine-Laurent Lavoisier

Diderot publishes the *Encyclopédie*

Seven Years' War

Joseph Priestley publishes *Experiments and Observations on the Different Kinds of Air*

American Revolution

THE ENLIGHTENMENT AND ENTERPRISE

6

I n 1727 François-Marie Arouet (1694–1778) witnessed the opulent state funeral of Sir Isaac Newton and reported to the intellectual world the status and importance of science in English society. Arouet, better known as Voltaire, was one of the leading lights of the French philosophical and social reform movement and was well placed to encourage the dissemination of Newtonian science to eighteenth-century continental circles. Voltaire and many other reformers were deeply influenced by the work of Newton and by the freedom of the English political system and sought to bring the best of rational thought from natural philosophy to the question of human relations. Voltaire wrote about the freedom of the English in his *Philosophical Letters on the English* (1734) and about Newton in his *Elements of Philosophy of Newton* (1738), written with Emilie de Breteuil, the Marquise du Châtelet-Lomont (1706–49). For many Enlightenment thinkers natural philosophy was both a model for, and a tool of, reform.

The spirit of reform worked in both directions. Just as the *philosophes* hoped to transform society, the new natural philosophers hoped to reform science, revolutionizing the practice of investigation, changing the language of scientific discourse, and placing their discoveries and expertise at the service of the state. Many natural philosophers took their research agenda and methodology from Newton, and so the mathematical and physical sciences of the eighteenth century, especially those concerned with force and matter, owed much to the venerable English scientist.

The eighteenth century was a time of burgeoning mercantile growth, as many European nations expanded their trade empires and colonies. With this expansion,

Europeans encountered cultures and peoples with different knowledge and understanding of the world, and so this period marks an important moment of the circulation of knowledge from the wider world back to Europe, as well as bringing European ideas of science and philosophy to other countries. Europeans were interested in products made from newly discovered flora and fauna, since such products could be very profitable in this new commercial revolution. It is, therefore, no surprise that many of those interested in studying nature increasingly did so in the service of commercial ventures or the state. Thus, two contradictory elements of scientific study developed in this period. On the one hand, the philosophical issues of freedom, democracy, and toleration led many natural philosophers to espouse radical political positions. On the other, scientists increasingly took their directions from the state or commerce and added their expertise to the exploitation of worldwide resources, with the result that they exerted a greater and greater influence on the wealth of nations and the power of the state. Many of the more radical efforts were overshadowed by the violence at the end of the century as reason fell to tyranny during the French Revolution, but by the nineteenth century science was a powerful and necessary ingredient for the operation of the modern state.

Universal Laws and Progress

The definition of the Enlightenment has been as hotly contested by historians as that of the scientific revolution. Different countries experienced different degrees of reform and resistance to reform. Still, at the heart of the Enlightenment were two monumental concepts. The first was a reappraisal of the human condition that led to the conception of universal human rights. The second was a belief in the inevitability of progress. Both led to cries for social, economic, and governmental reforms, and both owed much to conceptual changes in natural philosophy. One of the most obvious Enlightenment debts to the New Science, particularly following the Newtonian approach, was the belief in an underlying universality of natural laws. Newton had demonstrated that the law of gravitation and the laws of motion united the universe. There were no privileged realms and no exceptions; the laws of nature were the same for kings and peasants. It followed that similar universal laws existed and could be found to govern human relations. A number of philosophers argued for universal laws of human interaction in law, government, economics, social life, and religion. They were more than social critics; they called for action to amend the errors of the past. Like Francis Bacon, they challenged the very concept of traditional authority and hoped to replace it with rule based on reason and universal laws of human rights.

The philosophical tradition that led to the Enlightenment took shape when Newton was still alive. Thomas Hobbes (1588–1679) debated the nature of scientific knowledge with Robert Boyle and argued that a sovereign was necessary to protect people from each other and that to gain this security people had to give up personal rights. Boyle's view on scientific method won, while Hobbes's political views made him someone dangerous to know. He may have been rejected for membership in the Royal Society because of his political stance. John Locke (1632–1704) also refuted Hobbes's position on government by arguing for a natural and unalienable right to life, liberty, and the enjoyment of property. In his famous work *Two Treatises of Government* (1690), Locke defended the rights of the people to reject unwanted government, arguing that it was the duty of government to protect the rights of people and that the people had a right or even duty to depose any government that failed to protect their inherent and universal rights.

Historians have now cast doubt on whether Locke had much of an immediate impact on the course of English politics in the seventeenth century, but his ideas gained currency in the eighteenth century as more philosophers sought fundamental structures in society that paralleled Newton's structure of the universe. When Adam Smith (1723–90) wrote on economics and social interaction, particularly in *The Wealth of Nations* (1776), his analysis was based on laws of the marketplace that, like the Newtonian universe, were self-regulating. The "invisible hand" of the marketplace was not a description of a ghostly spirit of commerce but a mechanistic model of economics that he hoped would be as reliable and certain as the Newtonian model of the solar system.

A more subtle debt that the philosophers of the Enlightenment owed to natural philosophy was a belief in progress. At its simplest level, this was a belief that tomorrow would be better than today. Leibniz's formulation of this optimistic view was ridiculed by Voltaire, in his novel *Candide*, as that of a pedant believing in "the best of all possible worlds." However, the true natural philosophical articulation of this concept was much more complex. Rather than a vague belief in a better world, natural philosophers believed that they would understand the rules that governed nature better in the future than they did in the present. In fact, they believed that they would know everything about nature, identifying all the components of the machinery of the universe just as a clockmaker could identify and reassemble all the parts of a clock. This image of a clock fit well with Newtonian mechanics while at the same time was indicative of the growing power of industry, which could manufacture the precise and delicate mechanism of the mantel clock and the naval chronometer.

Diderot's *Encyclopédie*

The most ambitious and far-reaching intellectual project of the Enlightenment was Denis Diderot's (1713–84) *Encyclopédie*. When the first volume was published in 1751, it caused a sensation. French censors banned the book and revoked the publisher's license. France's attorney general called it a conspiracy against public morals, and the pope issued a declaration that anyone buying or reading the work would be excommunicated. The outcry from authority was so strong that it led some of Diderot's contributors to drop out of the project. In fact, his co-editor Jean d'Alembert (1717–83), a physicist who had worked on celestial mechanics, resigned over the controversy. There is nothing like a storm of controversy and whiff of scandal, however, to increase the desirability of such a work. Because of the furor, the *Encyclopédie* was soon the talk of Europe and reached a wide audience. Diderot had worked on, often under difficult circumstances, a monumental work that eventually reached 28 volumes, with 17 volumes of text appearing by 1765 and 11 volumes of illustrations finishing the series in 1772. It sold over 20,000 full sets, with thousands of abridged, partial, and pirated volumes in circulation as well.

Diderot's aim was nothing less than providing to all educated Europeans the most current information on every subject of modern thought. The spirit of the *Encyclopédie* was one of universal knowledge and a belief in progress; many of the articles contained the latest understanding of science, technology, and philosophy, complete with wonderful illustrations. Diderot believed that by making this knowledge available, the whole of civil society could be transformed. Just as the humanists had believed that education, based on learning the classics, made a person a good citizen, Diderot held that understanding science and the new knowledge it revealed would create a just society. The *Encyclopédie* was infused with social criticism, the promotion of modern industry, and a faith in the power of scientific discovery to bring enlightenment to society. It was both a compendium of the state of arts, crafts, and letters and a bold call for reform and progress. It also suggested order in nature and order in the study of nature.

6.1 ILLUSTRATION SHOWING A CHEMICAL LABORATORY FROM DIDEROT'S *ENCYCLOPÉDIE* (1772)

Scientific Societies and the Popularization of Scientific Study

The age of Enlightenment was also the period when natural philosophy began to flow into the general culture beyond a small elite of the intellectual class. The courtly philosophers had long provided a connection between natural philosophers and the seats of power, but with a few exceptions these connections existed only in aristocratic and elite social circles. The founding of the Royal Society in London, however, and the Paris Académie des Sciences had created a space for natural philosophy outside the university, the church, or the courts. With a declining number of princely courts in the eighteenth century, the creation of this new site for natural philosophical discussion and inquiry offered both a market and a reservoir of talent for an increasingly popularized scientific enterprise.

The two major scientific societies were copied across Europe, so by the end of the eighteenth century there was a scientific society in almost every major city and official national bodies in most countries. (See figure 6.2.) On the advice of Leibniz, the Prussian Academy of Sciences was created in 1700, making it one of the oldest in Europe. The Austrian Academy of Sciences began in 1713, while the Russian Academy of Science was founded by Peter the Great in St. Petersburg in 1724. The Royal Danish Academy of Sciences and Letters traces its roots back to 1742, and the Imperial and Royal Academy of Sciences and Letters of Brussels was founded

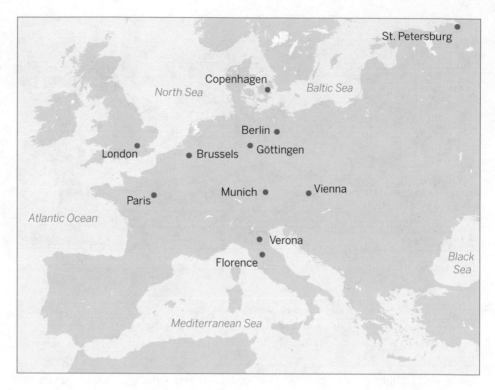

6.2 SCIENTIFIC SOCIETIES IN EUROPE, 1660–1800

during Austrian rule in 1772. A number of German societies including the Bavarian Academy of Sciences and Humanities (1759) and the Academy of Sciences of Göttingen (1751) were created mid-century. Italian societies included the Tuscan Academy of Sciences and Letters established in 1735 and the National Academy of Sciences (known as The Forty), founded in Verona in 1782.

The intersection of Enlightenment ideology, revolution, and science was particularly evident in the founding in North America of the American Philosophical Society in 1743 by Benjamin Franklin (1706–90) and others, including George Washington and Thomas Jefferson. The American Academy of Arts and Sciences, founded in 1780, counted among its founders John Adams and John Hancock and later listed Benjamin Franklin, John J. Audubon, and Louis Agassiz among its members. The connection between the reform of natural philosophy through these new scientific organizations and the reform of government by the men soon to be the founding fathers of the new American republic was integral to their overall philosophy and attitudes.

Due to rising levels of literacy and the spread of Enlightenment ideology, natural philosophy became a much more fashionable subject. In Britain, the locales of scientific discourse were the coffee houses and lecture halls, gathering places for the growing urban middle classes, at first in London and later in the provincial cities spawned by industrialization. This new audience was fascinated by the information natural philosophers provided, especially by the message of progress and the utility of such knowledge. Demonstrators like Jean Desaguliers (1683–1784) told people about Newton and his mechanics and exhibited new scientific principles using machines and models, which were closely tied to similar industrial innovations. These demonstrations and lectures provided not only a polite and rational language with which to think about the transformations taking place in society, both political and, especially, technological, but were also a morally acceptable diversion for both men and women. For religious dissenters, like Unitarians and Quakers, these introductions to God's rational power showed the utility of such knowledge and spurred the growth of science teaching at dissenter colleges, an important source of science education by the end of the century.

In France, this same interest in natural philosophy as a subject of polite and rational discourse was encouraged by the development of salon culture. The salons mimicked certain aspects of the princely courts but included people of a much wider range of backgrounds. They were often created by influential members of society and brought together artists, writers, musicians, and philosophers with the elites of commerce and government. The intellectuals were invited both to entertain and to inform. As such, the salons were also useful as conduits for patronage, especially for the bright and ambitious young men who lacked the resources or titles necessary to attend the king's court. It was for this audience that Voltaire's friend Count Francesco Algarotti (1712–64) wrote *Il Newtonianismo per le dame* (*Newtonianism for the Ladies*), published in 1735. Although Voltaire is best remembered today for his political and moral writings, he and his partner Émilie de Breteuil, Marquis du Châtelet (1706–49) published the *Éléments de la philosophie de Newton* (1738). Du Châtelet's translation of Newton's *Principia* published in 1759 with commentary and a derivation of the idea of the conservation of energy, remains the most important French version of Newton's seminal work.

One of the leading salons of the period was presided over by Madame Geoffrin. Born Marie Thérèse Rodet (1699–1777), she had married a wealthy husband and, particularly after his death, used her wealth and home to bring together interesting and important people. Madame Geoffrin was a close friend of Montesquieu, corresponded with Catherine the Great of Russia, and entertained both Benjamin Franklin

CONNECTIONS

Science and the Revolutionary Spirit

The Philosophical Society was founded at Philadelphia in 1743, making it the oldest learned society in North America. The founders included Benjamin Franklin, William Alexander, and Francis Hopkinson, all of whom would be major figures in the American Revolution. Franklin said of the reason for creating the Society that "The first drudgery of settling new colonies is now pretty well over and there are many in every province in circumstances that set them at ease, and afford leisure to cultivate the finer arts, and improve the common stock of knowledge."* In 1769 the Philosophical Society joined with the American Society for Promoting Useful Knowledge and was renamed the American Philosophical Society Held at Philadelphia for Promoting Useful Knowledge, which was quickly shortened to the APS.

What was established as a forum for people to discuss ideas and discoveries about nature turned out to be home to some of the most radical and revolutionary thinkers of the era. The connection between the study of nature and a desire for rationality and order in the human world proved to be a strong motivation to action.

The list of early members reads like a roll call of the Revolution, with George Washington, John Adams, Thomas Jefferson, Alexander Hamilton, Thomas Paine, and James Madison all joining. What these people had in common was a belief in reform and an abiding interest in natural

and Thomas Jefferson. She was a friend to Diderot and d'Alembert and supported the *Encyclopédie* financially and socially. Most of the intellectuals of the day who lived in or visited Paris attended her regular Monday or Wednesday meetings.

Electricity

A good example of the interest in natural philosophy evident in salon culture in France and natural philosophical demonstrations in Britain was the fascination with electricity. The existence of this mysterious substance had been known for generations, since William Gilbert identified it in the sixteenth century, but it was not easily subjected to investigation. Otto von Guericke in the seventeenth century had created a machine that generated an electrical charge by spinning a ball of philosophy and reason. To the members of the APS, reason and natural philosophy, particularly following the Newtonian model as presented in the *Principia* and in the more practical *Opticks* (both of which Franklin and Jefferson are known to have owned), were the basis for reliable knowledge. Just as nature functions under a single, universal physics, so too should human law be based on a single, universal law. Paine, for example, argued that society existed in a state of "natural liberty" (meaning that human rights originated in nature, not human convention) in which all people were equal, not divided into higher and lower classes. This was very much in the Enlightenment tradition and it was made manifest in the documents of the American Revolution. This is especially true of the Declaration of Independence (the writers of which included members of the APS), which includes direct reference to the laws of nature in the introduction:

> When in the Course of human events, it becomes necessary for one people to dissolve the political bands which have connected them

with another, and to assume among the powers of the Earth, the separate and equal station to which the Laws of Nature and of Nature's God entitle them, a decent respect to the opinions of mankind requires that they should declare the causes which impel them to the separation.

The Preamble presents human rights as axiomatic just as physical laws are axiomatic and states that there should be one universal, natural, and equal law for all people, just as Newton's law of gravity was universally applicable. Although natural philosophy was not the cause of the American Revolution, it was part of its intellectual heritage. The APS would continue to foster revolutionary ideas, including among its membership Charles Darwin, Louis Pasteur, Elizabeth Agassiz, Linus Pauling, and Margaret Mead.

* Benjamin Franklin, "A Proposal for Promoting Useful Knowledge among the British Plantations" (original 1743, *American Philosophical Society*, 1768), www.amphilsoc.org/item-detail /franklins-proposal.

sulfur and rubbing it with the hands or a cloth. This allowed for a controlled production of a static charge but was still of limited use as an experimental device. In the 1750s a number of people designed generators that used spinning glass disks to generate the charge, the ultimate example being a massive device built by John Cuthbertson (1743–1821) that could generate a charge of about 500,000 volts.

Smaller static generators were frequently used for demonstrations such as the one featuring an electrified boy. In it, a child was suspended from the ceiling by silk cords, a charge was generated and communicated to him, and he then moved small objects using the static electricity or delivered mild shocks to visitors. The electrified boy was clearly an entertainment, as were games based on passing a shock from person to person or electrifying chairs to zap unsuspecting visitors. (See figure 6.3) But there was a point to such demonstrations: they were often used to illustrate concepts such as electrical flow, circuits, and insulation.

6.3 THE ELECTRIFIED BOY

Charged with static electricity (rod at "n") and suspended by silk cords, the boy attracted bits of paper to demonstrate the communication of electricity. From Johann Gabriel Doppelmayr, *Neu-entdeckte Phaenomena von bewunderswürdigen Würckungen der Natur* (1774).

The initial research on electricity was qualitative, concentrating on finding electricity, observing its behavior, and learning how to manufacture it. Franklin's 1752 kite experiment is the most famous episode in the hunt for electrical phenomena and has risen to the status of popular mythology. It is likely that Franklin did fly a kite as an experiment to determine whether lightning was electricity, but he left no direct account of the event nor any indication of the date it might have taken place. The clearest record comes from Joseph Priestley's reporting of Franklin's account, but this was published 15 years after the fact. What is certain is that Franklin did not fly a kite in the middle of a thunderstorm as is popularly depicted. He flew a kite at the approach of storm clouds, reasoning that the accumulation of electricity must precede the discharge of lightning. He broke the path of the electricity by attaching the kite rope to one side of the loop on top of a key and a silk ribbon to the other side, which he was careful to keep dry (and thus insulated). He ran an iron wire from the key to the ground. When he noticed the fibers of the hemp twine of the kite rope stand up and repel each other, he put his knuckle near the key and got a shock. While this was convincing evidence, he also used the wire to charge a Leyden jar to demonstrate that lightning charge was identical to that collected by conventional generation.

While qualitative work was a necessary first step in understanding electricity, it was of limited value in understanding how electricity worked or what could be done with it. In particular, the problem of complete discharge made electricity difficult to study. The creation of the Leyden jar around 1745 was a major step toward controlling the mysterious energy. It was not discovered by one person but by three at almost the same moment: Ewald Jürgen Kleist (1700–48) in Germany and Pieter van Musschenbroek (1692–1761) and Andreas Cunaeus (1712–88) in Holland. While Kleist was probably the first to use a jar filled with water to try to collect an electrical charge, it was Musschenbroek who described the instrument. Since the work was conducted in Leyden, over time that was the name adopted by most people (although it is still sometimes called the *von Kleistiche Flasche* in Germany). A glass jar was wrapped in metal foil (this was thought to prevent the electricity leaking through the glass) and filled with water. A spike or metal rod pierced a stopper that sealed the jar. Electricity from a static generator was conducted to the interior through the rod and stored. When the rod touched a conductor, the electricity was discharged. (See figure 6.4.)

6.4 LEYDEN JAR

Franklin used the Leyden jar to explore the behavior of electricity in a very practical manner. He used the discharge to mimic lightning and designed the lightning rod to protect buildings. He made a model called the Thunder House to demonstrate the effectiveness of the lightning rod. Inside the model house was a small amount of flammable gas in a closed container. When electricity was applied to the top of the house, it produced a spark that ignited the gas and blew the lid off the container and the roof off the house. With the lightning rod attached to the house, the charge was conducted away from the model and the house was safe from damage. Franklin was later called to London to discuss the effectiveness of lightning rods with King George III. The king remained convinced that such rods were more effective with a round globe at the top, and this rather inefficient design was adopted by the British.

Theory of the period described electricity as a fluid, which in effect "flowed" into the Leyden jar from the electrostatic machine. The jars also helped to clarify the concept of an electric circuit, and it was from this that another salon demonstration arose. After filling a Leyden jar, a circle of people joined hands. One person broke the circle and with the free hand held the bottom of the jar, while the person on the other side of the break touched the rod with their free hand. If there was sufficient charge, everyone in the circle felt a shock. Because the Leyden jar could hold only a fixed amount of electricity (its capacity; hence, the name capacitors for devices based on the concept), such demonstrations helped to shift the study of electricity from a qualitative examination to one based on quantitative ideas, such as the relationship between the area of coated surface, the amount of charge, and its force.

New Systems of Measurement

The difficulty of doing experiments on electricity highlights one of the central problems for scientific research in the period. Electric phenomena could largely be understood qualitatively but not quantitatively. The precision of experimental results lagged far behind the potential revealed by physics and mathematics. The location of distant planets could be more accurately determined than the place of a ship at sea. The increased interest in quantification, developed from the seventeenth-century allegiance to measurement as the path to true knowledge, led in the eighteenth century to the creation of new instruments and, in keeping with the age of reform, the introduction of a new system of measurement.

Chief among these new instruments was the thermometer. Although there had been a number of attempts to measure temperature, including work by Galileo, the first instrument accurate enough for scientific work was not created until 1714 when Gabriel Fahrenheit (1686–1736) used the expansion of alcohol to measure temperature. He also introduced the use of mercury as the expanding liquid and published his method of construction in the *Philosophical Transactions of the Royal Society* in 1724. He went on to produce a scale based on the freezing points of water, establishing 0° as the freezing point of salt water, 32° as the freezing point of fresh water, and 212° as the boiling point.

Setting zero as the freezing point of salt water made sense in an era when understanding the sea was crucial to trade and power. In the same year that Fahrenheit created his first mercury thermometer, the English Parliament through the Board of Longitude offered a reward of £20,000 (equal to about $500,000 today) for anyone who created a system for determining longitude at sea. This was a long-standing problem that had plagued navigators for generations. Both the Paris Observatory (created in 1667) and the Royal Observatory at Greenwich (established in 1675) had the improvement of navigation as one of their main projects. Longitude could be calculated from celestial observations on land but not on shipboard. Over the years a number of solutions had been proposed, but all had failed as either impossible to carry out on a moving ship, such as the observation of the moons of Jupiter, or simply irrelevant, such as measuring magnetic variation. What prompted the 1714 prize was the ongoing danger of navigation, now more a concern than ever due to the expansion of trade and volume of shipping. The necessity of better navigation was made very clear when in 1707 Admiral Sir Cloudesley Shovell lost four ships and died along with between 800 and 2,000 men in a shipwreck off the coast of England. While finding latitude was relatively easy using celestial observations, what was needed to find longitude was an accurate method of timekeeping. This had been understood since at least 1530 when Gemma Frisius (1508–55), mathematical geographer and teacher of Gerardus Mercator, argued that he could determine longitude if he had an accurate clock.

Therein lay the problem. Measurements of time accurate enough to determine longitude were possible using astronomical events such as the eclipses of the moons of Jupiter, as suggested by Galileo, but such observations were impossible on a ship at sea. Even after Christiaan Huygens developed a pendulum clock in 1656 that was accurate enough to keep good time, the instrument was useless on board moving ships. The problem was finally solved by the clockmaker John Harrison (1693–1776), who in 1761 had his timepiece tested on two voyages to Jamaica. Despite the success of Harrison's chronometer, he was awarded only £5,000 by the Board and did not receive the whole prize until King George III interceded on his behalf in 1773. As a craftsman he did not fit the gentlemanly profile of natural philosophers developed by the Royal Society. The leadership of the Royal Society had expected one of their Fellows would win the prize.

Navigation was vital to European nations competing for colonial holdings and operating global trade networks. This race to empire dictated the subject of some research programs, like that for longitude and the cure for scurvy. At the same time, the expanding reach of these nations made possible a number of experiments and observations that could have been conceptualized but not undertaken in earlier times. Take, for instance, the expedition to Lapland of Pierre de Maupertuis (1698–1759). Maupertuis was a member of the Académie des Sciences and in 1736 was chosen to lead the expedition to measure the precise length of a meridian degree. He was a fervent Newtonian and had been one of the first to introduce Newton's theory of gravity to French scientists. The resulting report, *Sur la figure de la Terre* (*On the Shape of the Earth*, 1736), confirmed Newton's theoretical prediction that the Earth was not spherical but flattened at the poles.

Mathematical Physics

Maupertuis was one of a growing group of continental thinkers who turned their work to mathematical physics. They bridged the disciplines of mathematics and experimentalism, often following the direction set by Newton both mathematically and in practical terms by the research suggested in the "Queries" section of Newton's *Opticks*. Among this group were Joseph Lagrange (1736–1813), Pierre-Simon Laplace (1749–1827), Augustin Jean Fresnel (1788–1827), and Leonhard Euler (1707–83), one of the most prolific mathematicians of all time, who published over 1,000 articles and books on everything from number theory to mathematical cartography. Although the interests and studies of these men were diverse, ranging from the most esoteric aspects of the calculus to chemistry, what ties their work together was their effort to put into mathematical terms everything that was around them. Fresnel worked out a wave theory of light, challenging the corpuscular theory of the Newtonians, but he also developed the Fresnel lens, which was built in a series of concentric rings. This was used in lighthouses (and later in theater lighting) because it allowed a much greater concentration of light than traditional metal reflectors had permitted, but was considerably more compact than a regular lens.

Lagrange began his serious mathematical work at the age of 17 by looking at optics. He sent a proof of his results to Euler, who encouraged him to continue in this endeavor. Lagrange did continue and essentially founded the field of astrophysics. While Newton's mechanics had set a framework, the precise details about why the solar system moved as it did or the nature of the gravitational relations of the Earth and its Moon were complex problems that needed to be worked out. Lagrange did so in his book *Mécanique analytique* (1788), which presented mechanics in strict mathematical terms. The Preface stated:

> One will not find figures in this work. The methods that I expound require neither constructions, nor geometrical or mechanical arguments, but only algebraic operations, subject to a regular and uniform course.[1]

Lagrange's work on gravity led him to predict that there would be five points in space – the Lagrange points – where the gravitational forces of two bodies (such as the Earth and the Moon) balanced each other. Of the five points, two (L_4 and L_5) were more stable than the others, and the forces of attraction worked in such a

1. Joseph Lagrange, "Preface," *Mécanique analytique* (Paris: Gauthier-Villars et fils, 1889), 1.

way that a third body placed there could remain in stable orbit indefinitely. The L_1 spot in the Earth-Sun system is currently occupied by the Solar and Heliospheric Observatory Satellite, which is, as the name implies, observing the Sun. (See figure 6.5.)

Like Lagrange, Laplace also worked on celestial mechanics. He wrote many technical papers, as well as *Exposition du Système du Monde* (1796), a popular account of how the universe functioned. In practical matters, he was an assistant to the great chemist Antoine Lavoisier, aiding him in his work on heat and energy. These scientists added new vigor to physics, turning their attention to a wider and wider range of natural phenomena. One of the grandest schemes was offered by Roger Joseph Boscovich (1711–87), who claimed that all matter in the universe could fit inside a nutshell and who tried to encapsulate the structure of the universe in a single physico-mathematic idea. His *Theoria Philosophiae Naturalis* (1758) eliminated matter altogether and dealt only with points of force, presenting one of the first attempts at a grand unified theory (such theories are sometimes called GUTs) based solely on physical principles.

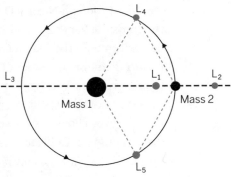

6.5 LAGRANGE POINTS

The Lagrange points (L_1, L_2 etc.) are points in space where satellites would have the most stable orbits because the gravitational forces are balanced between Mass 1 and Mass 2.

Scientific Expeditions: The Transit of Venus

Grand mathematics stirred the hearts and minds of a number of scientists, but it was often the adventures of investigation that attracted public notice. Among the most famous scientific expeditions of the period were the investigations of the transit of Venus. Kepler had earlier recognized that both Mercury and Venus must pass in front of the Sun and had calculated when these events would occur. He determined that the transit of Venus happened in pairs every 105.5 years (with the second eight years later), followed by an interval of 121.5 years (and another eight years). Just a year after Kepler's death, Pierre Gassendi had observed the transit of Mercury in 1631. It was now time to observe the next transit of Venus, due in 1761 and 1769. The importance of this astronomical event was recognized by Jeremiah Horrocks (1619–41), who realized that if observations were made simultaneously at different points on the Earth, the information gained could be used to calculate the distance to Venus and the distance from the Earth to the Sun.

In 1761 Joseph-Nicolas Delisle (1688–1768) organized a major effort to measure the transit of Venus by sending observers to Pondicherry in India, St. Helena, Isle de Rodrigues in the Indian Ocean, and Tobolsk, Siberia. Observations were undertaken by many other observers as well, but unfortunately poor weather and the Seven Years' War (1756–63) prevented some of the observers from recording the event, and the results were not as good as astronomers had hoped.

The expeditions of 1769 were more successful. Captain Cook's voyage to Tahiti and the adventures of Guillaume Le Gentil (1725–92) captured the public's romantic imagination. Tahiti had only recently been discovered by British explorers and was portrayed as an earthly paradise. Because it had a superb tropical climate and the islanders were extremely friendly, reports about Tahiti were widely spread by the European press. Cook's mission was in large part scientific – mapping, observing the transit, testing the chronometer for longitude, and looking for a way to deal with scurvy – but he was also under orders to claim any previously undiscovered land for the British.

Le Gentil went to Pondicherry, undertaking a difficult trip overland and then across the Indian Ocean, but missed the first transit of Venus. He stayed in the region until the second transit, only to be thwarted again because of bad weather. He wrote of his adventures in *A Voyage in the Indian Ocean* (Volume I published in 1779, Volume II in 1781). So began the age of heroic science, when facing extreme danger in the name of the expansion of knowledge gained a significance almost equal to the religious crusades of long ago. The Académie des Sciences team, sent to Mission San Jose del Cabo in Baja, California, all died from an epidemic except for a single survivor. This man crossed the territory that would become Texas on foot in grueling conditions and returned to French territory and home only after great travail. It took several years to compile the data (Cook, for example, did not get back to Britain until 1771, since he was searching for the Great Southern continent), but the astronomer Jerome de Lalande (1732–1807), the director of the Paris Observatory, used the combined transit data from all these teams to calculate a distance of 153 to 154 million kilometers from the Earth to the Sun. While this was not as accurate as astronomers had hoped, it was a major accomplishment. It also demonstrated how scientific knowledge, as well as guns and colonies, confirmed the imperial power of European nations such as France and Britain.

Other Scientific Encounters

Europeans were interested in discovering more about the vast resources of other parts of the world, for commercial, political, and scientific reasons. This was the period that saw the first concerted efforts to map countries and regions largely unknown to Europeans. For example, the first British maps of the subcontinent of India were made in the eighteenth century. James Rennell (1742–1830) produced maps of the Mughal Empire, relying on assembling in London the diverse field reporting of members of the East India Company. Europeans were also interested in the uses of plants and animals in other regions and were reliant on native knowledge to give them this information. Maria Merian's investigation of the properties of insects in Suriname was only possible because the Surinamese shared their knowledge with Merian. Europeans and non-Europeans were often on an equal footing when it came to knowledge of the natural world, although information not seen as useful to Europeans was frequently not transmitted home. For example, knowledge of the abortificant (abortion-causing) properties of the peacock flower of South America were not communicated to Europe, although the flower itself became known and prized for its beauty.

Perhaps the event that best demonstrates this negotiation between scientific worldviews was the diplomatic mission of George McCartney (1735–1806) to China in 1793. McCartney was sent by the British to establish better trading relations with the Qing dynasty. He brought diplomats and soldiers, but he also brought scientists and scholars, along with scientific devices he felt would astonish the Chinese (for example, a world map and an ornate clock). Joseph Banks advised him to find as many tea plants as possible, with an eye to diversifying the new tea plantations Banks had already founded in India in 1778. The mission was a complete failure. McCartney got into a difficult situation because he refused to kowtow to the Qianlong emperor, and the Chinese were not impressed by these marks of European scientific knowhow. They had their own maps of the world and their own time-keeping devices. The British had expected to encounter a backward people, eager for instruction. Instead, they met a sophisticated and educated culture, with its own scientific understanding.

The Industrial Revolution and the Study of the Earth

Geographers, mathematicians, cartographers, astronomers, and clockmakers all worked to solve the complex problems of navigation and astronomy, thereby contributing to the power of their nascent empires. Just as importantly, their success aided the exploitation of the natural resources and captive markets of the colonies. The wealth so generated in turn helped finance the Industrial Revolution, particularly in Britain, where there was a burst of creative industrial activity

starting around 1750. By 1780 the introduction of the factory system of production
and James Watt's (1736–1819) condensing steam engine had begun to transform
the lives of many people as they moved from the farms to the factories, creating on
one hand the terrible slums of early industrial Glasgow, Birmingham, Manchester,
and London, and on the other the incredible wealth of the industrial barons.
Historians of science have long debated the relationship between the growth of
science and the technological and economic changes in this period. Rather than
scientific breakthroughs contributing directly to new technologies, however,
science was instrumental in creating a culture of progress and a discursive claim
for the utility of the scientific enterprise.

As merchants were sailing to and trading with new parts of the globe, natural
philosophers came to view the Earth itself as a legitimate topic of scientific
investigation. Studies of the Earth – cosmography and geography in the scientific
revolution, adding geology in this Enlightenment era – were motivated by economic considerations to do with mining and land formations, political issues
dealing with newly discovered peoples and the navigability of the globe, and
religious questions of the evidence for God's handiwork. All contributed to geology
and Earth studies in the eighteenth century, which were focused on several
questions: the process of creation and subsequent history of the Earth, the age of
the Earth, the configuration of strata, and the place of fossils.

Natural philosophers were first concerned with the material reality of God's
creation. The Englishman Thomas Burnet (c. 1635–1715), for example, in *Sacred
Theory of the Earth* (1691), wanted to explain creation in mechanical (Cartesian)
rather than miraculous terms, following the trend of mechanical philosophy to
avoid supernatural explanations. He started with the Genesis story but used
natural, mechanical explanations to account for the events of biblical creation. He
claimed that the Earth in the beginning was very hot; as it cooled, a thin shell of
smooth land formed above the waters. Burnet did not eschew all biblical explanations, however, since he argued that the coming of sin into the world caused the
shell to break, sending floods over the land to create the imperfectly spherical
world, with mountains and oceans.

Burnet was criticized for his irreligious stance, despite his use of biblical
reasoning. This did not prevent another English natural philosopher, William
Whiston (1667–1752), from making a similar point in *New Theory of the Earth*
(1696). Whiston was a Newtonian – he had been at Cambridge with Newton – and
sought to place Earth's history in the context of Newtonian thought. He argued
that the Earth was formed by a comet that condensed into a solid body because of

gravity and that the deluge was caused by another comet passing close by, depositing water, and knocking the Earth out of circular orbit. Hence, the imperfection after the Fall of Adam and Eve.

These late seventeenth-century creation stories were coupled to the heated religious conflicts of the age. In the eighteenth century it was more possible to move away from biblical stories and look instead for rational explanations. This is not to say that eighteenth-century geologists were atheists, however. Rather, they saw God's work as essentially rational and therefore explicable. As well, the rise of the amateur gentlemanly philosopher brought a new purpose to geological investigation – for status and political power rather than for ideological or religious reasons. Geology, for instance, concentrated on the observation of actual phenomena, including fossils, rather than on large speculative systems, such as those of Burnet and Whiston. In the early seventeenth century, fossils were considered to be figured stones rather than biological in origin, but now scientists such as Nicolaus Steno (1638–86) and John Woodward (1665–1728) interpreted them as the remains of living creatures that had been petrified in rocks.

Recognizing fossils as biological remains, however, caused more problems than it solved. Most of the fossils found in Europe seemed to be of sea creatures and did not resemble contemporary living forms. Moreover, many of the best fossil deposits were high up on mountains, where people could not believe there had been an ocean or lake at an earlier time. While the biblical flood might provide some explanatory options, extinction and submerged mountains required a major leap of scientific faith. This crisis of explanation led to the development of two competing geological theories by the end of the eighteenth century: Neptunism and Vulcanism, named after the Roman gods of the seas and fire.

Neptunism was so named because its proponents believed that water was the fundamental agent for the formation of the Earth. George Louis Leclerc, Comte de Buffon (1707–88), argued in his *Histoire Naturelle* (1749) that a cooling Earth had gone through six epochs of earth formation, corresponding roughly to the six days of creation. In these six epochs, the Earth cooled, water condensed, and oceans retreated; the Earth evolved in a clear direction toward the present day. The leading voice of Neptunism, however, was Abraham Gottlob Werner (1749–1817), which is why this theory is sometimes called the Wernerian theory. Werner was a mineralogist teaching at the Freiberg Mining Academy, one of the new educational innovations of the Enlightenment. He decided that strata must indicate the order of rock formation and that the layers had been precipitated out of primordial oceans. His explanation was based largely on his findings of sedimentary rock formations;

he assumed that volcanic activity was local and relatively unimportant. He was probably influenced in his formulation by the lack of such volcanic activity around Freiberg, where he worked. Werner also insisted that forces now in action were too feeble to have made the world as it now is, and so forces in the past must have been far stronger and more powerful. In other words, he believed that geological formation operated in only one direction, toward the present. Werner was extremely influential, because he taught many mineralogists, miners, and geologists, and because his scheme fit with a classification scheme for minerals. It also appealed to the growing number of researchers with interests in chemistry, since it suggested a chemical foundation for the different types of rocks and minerals.

Neptunism was widely popular until the end of the eighteenth century, but it could not explain a number of geological observations such as where had all the Flood water gone? How were igneous rocks formed? And, the most vexing question, why were fossils of sea creatures found on the top of mountains? Gradually, a number of mineralogists began to argue for some contribution of the inner heat of the Earth, sometimes called Plutonism or Vulcanism. The idea that heat, specifically in the form of volcanic activity, accounted for geological structures had been suggested as early as 1740 by Anton Lazzaro Moro (1687–1764) from his study of volcanic islands. He also argued that aquatic fossils found on mountains meant that the rocks here were once at the bottom of an ocean. The most dedicated vulcanist was Jean Étienne Guettard (1715–86). Guettard lived in France near Auvergne where there were a number of mountains with cone shapes, indicating they had once been volcanoes. He argued that volcanic forces in the past had been much stronger than now and widespread enough to account for rock formation, fossil placement, and igneous rocks. He was the first person to create geological maps of France, his *Atlas et description minéralogiques de la France* (*Mineralogical Atlas and Description of France*) appearing in 1780. Along with a young Lavoisier, he contributed to a project to do a complete geological survey of France, of which only a tiny portion was completed. Both Guettard and Werner were important enough as scientific thinkers to have been elected as foreign members of the Royal Swedish Academy of Sciences; Guettard being elected in 1759 and Werner in 1810.

Although Guettard and Werner represented the most distinct versions of Neptunism and Plutonism, the more people began to investigate rocks and their layers, the less satisfying both these accounts seemed to be. James Hutton (1726–97), a Scottish geologist, developed a new theory that combined the inner heat of the Earth with the actions of water. In *The Theory of the Earth* (1795), he suggested that sedimentary rock had been formed from earlier igneous rock and that the Earth was in a constant state of change from one form into the other and back again. Volcanic activity forced igneous rocks up to the surface, where water caused erosion and the formation of sedimentary rocks, some of which were fused by heat to form igneous. And so the cycle went. As Hutton said, "We find no vestige of a beginning, no prospect of an end."[2]

With this statement, Hutton attacked all the mineralogists and geologists before him, since he argued that forces now in action did have sufficient strength to cause the evolving nature of rock stratification and that the inner heat of the Earth was the same now as before. In other words, this was a model *without direction*, either biblical or material. This is called "uniformitarianism." Hutton's was a wildly radical theory, and although his earlier papers were favorably reviewed when they first appeared in 1785, with the publication of his book in 1795 he was widely condemned. This was largely because the French Revolution in France had whipped up conservative concerns, and Hutton's "atheist" and "Jacobin" theories had no place in a patriotic country at war. The liberal legacy of the Enlightenment was over, at least for a time.

Museum Collections and Scientific Expeditions

The importance of geology in the eighteenth century was mirrored by the passion for collection and classification. Collectors in the sixteenth and seventeenth centuries had assembled their collections in museums, first as private places of study or contemplation (a *studio*). These early collections by natural philosophers such as Konrad Gesner (1516–65) and Ulisse Aldrovandi (1522–1605) included books, engravings, and artifacts. The selecting and organizing principles in these museums were usually based on the singularities, freaks, and discontinuities of nature rather than some underlying order or continuity. Collections might include stones in bizarre shapes or two-headed calves, rather than a complete set of the flora of a region. In the late seventeenth and early eighteenth centuries princes developed collections as part of the spectacle and wonder of courtly science. These *studiolo* or cabinets of curiosities (which were de rigueur for gentlemen by the nineteenth century) were not working collections, as Gesner's and Aldrovandi's had been, but rather were meant for public display and ostentation.

The first truly public museum, meaning that the public could pay a fee and tour the collection, was the Ashmolean Museum, established in Oxford in 1683. Its

---

2. James Hutton, "Theory of the Earth," *Transactions of the Royal Society of Edinburgh* 4 (1788): 304.

model in terms of public access was probably the Bodleian Library, also in Oxford, the first European public library, which people could enter for a fee if they had a reason to use it. Not all visitors were pleased with the open-door policy, since it violated the privileged status of the gentlemanly witness. Count von Uffenbach of Germany, for example, visited the Ashmolean in 1710 and was not at all impressed, since he was forced to rub shoulders with the hoi polloi! However, public museums, or at least museums that could be accessed by an educated public, soon became a mark of enlightened Europe. Museums in Russia, Italy, Spain, and France all opened their doors to the public in the eighteenth century (although opening the Jardin du Roi in France required the storming of the Bastille). The British Museum opened in 1753, following a bequest by the late president of the Royal Society, Hans Sloane (1660–1753), from his own collection.

Sloane, a physician to the wealthy in London, amassed an enormous collection. At first he collected Old and New World plants, and then branched out to include a huge array of other artifacts, including shells, insects, fossils, minerals, antiquities, coins, books, and manuscripts, most collected for their rarity rather than to create a coherent classification system. He began his collection, as well as his path to riches and power, on a trip to Jamaica in 1687 under the patronage of Christopher Monck, Duke of Albemarle. When the duke unexpectedly died on the voyage, Sloane had a somewhat grotesque opportunity to practice his preservation technique, embalming Albemarle's body to preserve it on the trip home. On his return Sloane made use of one of the new species he had identified in South America by marketing it to the English public. He had observed South American natives consuming chocolate but found it too bitter for his own taste. However, he discovered that mixed with sugar and milk, the result was very pleasant, and so he made his fortune importing and manufacturing milk chocolate.

Sloane's collection was visited by most of the important people of the day – Carolus Linnaeus, Benjamin Franklin, and Georg Handel, to name a few. Sloane himself was a somewhat touchy and antisocial man who got into prolonged fights with nearly every important scientist in early eighteenth-century Britain. For example, he had a huge row with the geologist Woodward, whom he accused of grimacing at him at a Royal Society meeting. In 1742 he retired to Chelsea, outside London, and set up his Physic Garden. At his death he established a trust that soon formed the British Museum, which became the largest public museum of the eighteenth century.

Sloane's successor as president of the Royal Society, and an extraordinary collector in his own right, was Sir Joseph Banks (1743–1820). Early in his career

First Voyage (1768–1771)
Second Voyage (1772–1775)
Third Voyage (1776–1779)

Atlantic
Ocean

Pacific
Ocean

Indian
Ocean

6.6 CAPTAIN JAMES COOK'S THREE VOYAGES (1768–1779)

Banks signed on as ship's botanist for James Cook's (1728–79) first voyage to
the Pacific in 1768. Indeed, on this first voyage, Banks, who represented the
Royal Society on the trip, was more famous than Cook, who was a relatively
unknown naval officer. Banks's entourage threatened to overwhelm the ship,
and, in fact, his demands for space and resources were so great that he was
not asked to participate in Cook's next voyage. Banks was overwhelmed by
the wealth of new flora and fauna he observed on this journey. He named
their first landfall in Australia Botany Bay, because the vegetation was so
luxurious and exotic. He returned determined to collect all of the world's
botanical species.

Cook's voyages themselves demonstrated an intersection of empire-building,
heroic science, and the collecting spirit. Cook set out to prove the existence or
nonexistence of *Terra Australis Incognita*, the theoretical continent that should
have balanced the Eurasian continent on the other side of the globe. He proved
that it did not exist, while at the same time discovering a number of new lands

Jesuit Science
Overseas,
1540–1773

From the founding of the Jesuit order by Ignatius of Loyola (1491–1556) in 1543 to its suppression in 1773 (for complex political reasons), Jesuit missionaries founded schools, colleges, universities, and missions all over the world. At its peak in 1750, the Jesuits operated over 500 colleges and universities in Europe, another 100 or so overseas (mostly in Spanish America), and had around 270 missions around the globe. The order frequently worked with and advanced the interests of the great colonial empires of Spain, Portugal, and France. At the same time, the Jesuits provided an international network for the circulation of natural knowledge around the world, interacting with local rulers, Indigenous peoples, colonial governments, and the intellectual and religious organizations of Europe. It was a truly international and intercultural organization.

Integral to the work of Jesuits was the investigation of nature. Jesuit fathers were well educated in mathematics and astronomy and interested in many different scientific areas, especially astronomy,

and islands (and claiming them for Britain), testing the chronometer, and establishing that eating limes could prevent scurvy. These voyages were symbolic of the British Empire, collecting and laying claim to parts of the world and establishing the power of British science and technology. Although Banks did not go on Cook's next two voyages and therefore did not witness Cook's murder on Hawaii nor travel to Vancouver Island, he did issue instructions, with the approval of the Admiralty, for the collection of flora and fauna and the observation of natural phenomena.

Banks became the greatest botanical collector in Europe, as well as the most powerful natural philosopher in Britain, controlling appointments and influencing patronage. He established Kew Gardens, not only collecting specimens but growing them and distributing them to other gardens. Most of the oriental flowering plants that are now an integral part of European horticulture – for example, rhododendrons – came from Banks's collection. Banks saw Kew Gardens as a great exchange house for the British Empire; he used it to help transport plants from one part of the Empire to another – with exempla, of course, remaining at Kew. Here, then, was museum-collecting with a new imperial thrust. As a cartography, geography, natural history, ethnography, botany, and *materia medica*. Their interests in nature took different forms in different places. For example, in China, from the time of the permanent residency of Matteo Ricci (1552–1610) in Beijing (from 1601), Jesuits became closely involved with the Astronomical Bureau in Beijing, working collaboratively with Chinese astronomers and scholars. From the arrival of French Jesuits in 1684, they were commissioned by the Qing emperor Kangxi (r. 1662–1723) to participate in a huge cartographical project to map the growing Chinese empire.

In most of their global outposts, on the other hand, Jesuits were interested in medical botany, both for practical reasons and as part of their interest in the wonders of the world that would show the greatness of God. In Spanish America, French Canada, and Portuguese West Indies, they gathered knowledge of plants and their cures from Indigenous people, translating such knowledge through letters and publications to Europe. They also sent specimens back to Europe, contributing to the great European collections of scholars such as Konrad Gesner. We would see this as an appropriation of native knowledge, only possible because Jesuits established long-term residency in these places, learned the languages and the cultures, and worked to gain the trust and support of Indigenous peoples and local rulers.

The Society of Jesus was restored in 1814, and several members of the order in later years continued to support the study of nature, both in Europe and internationally.

personal friend of George III, Banks was able to instruct expeditions of the Royal Navy to collect and transport specific samples. He also hired professional collectors, such as Archibald Menzies (1754–1842), who brought back the monkey puzzle tree and the giant redwood, and Mungo Parks (1771–1806), the famous African explorer, who eventually died in Africa trying to discover the route of the Niger. Banks's two most famous exploits were transplanting tea from China to India in 1788 and transplanting breadfruit from Tahiti to the West Indies in 1787. The latter employed the services of the HMS *Bounty*, and sailors impatient with poor conditions for the men while the breadfruit received good treatment mutinied against Captain Bligh and resettled on bleak Pitcairn Island. Perhaps the most astonishing part of the mutiny on the *Bounty* is that Bligh, with his supporters, succeeded in navigating back to Britain from the Pacific in an open boat and set out soon thereafter with a new ship to transplant the breadfruit, this time successfully.

Following the British example, other European nations strove to create an empire through scientific observation as well as military coercion. The French, particularly, made an imperial bid for the Pacific just as the British were doing. For example, Louis-Antoine de Bougainville, Comte de Bougainville (1729–1811), landed at Tahiti, brought back a new plant named in his honor, *Bougainvillaea*, and set in motion the philosophical paradigm of the noble savage, as articulated by Jean-Jacques Rousseau. Jean-François de Galaup La Perouse commanded another voyage of discovery, exploration, and empire building, but his ship was lost in the Pacific in 1787, leading to many romantic speculations as to his fate. Much time, imagination, and money were spent trying to find this lost expedition, but since the French Revolution broke out almost immediately, the first ship sent to search mutinied and returned home, and La Perouse was never heard from again.

Amateur Scientific Societies

For those who remained at home, the curiosity for collecting and understanding natural phenomena and for attending natural philosophical lectures, as well as the legitimating power of scientific knowledge, led to the formation of a number of amateur-based scientific societies. Perhaps the most famous was the Lunar Society of Birmingham.

The intersection of scientific utility with industrial and economic interest was illustrated by the collaboration of the members of the Lunar Society. Based in Birmingham, it started in 1765 when a small group of men met informally to discuss natural philosophy and issues of the day. They called themselves the Lunar Circle. The group expanded in 1775 at the instigation of William Small (1734–75) and Benjamin Franklin and was renamed the Lunar Society, as a reference to the fact that the members met monthly on the Sunday or Monday evening closest to the full Moon so there was more light by which to travel the unlit streets of Birmingham. Meetings of the "Lunatics," as they called themselves, often took place at Soho House, Matthew Boulton's home. The Society faded around 1790, largely because the French Revolution made them seem a potentially subversive group, but sporadic meetings may have continued until 1809.

While the Lunar Society was never as formal as the various scientific societies of the day – in fact, many of its members were also Fellows of the Royal Society (FRS) – it took the concept of putting the utility of scientific knowledge into practice far more aggressively. Knowledge of the natural world was not simply for a better understanding of God's creation or to improve the mind. Knowledge was to improve the human world. For the most part, these men saw no distinction between personal profit and social benefit. Many of the members were business-men first and natural philosophers second, and most worked in a wide and diverse range of occupations during their careers. While some aspects of reform were of interest to them, they might be better characterized as "improvers," restless and constantly tinkering and looking for ways to make things work better, whether it was an engine, a business practice, education, or the production of sulphuric acid. In particular, the collaboration of James Watt with the industrialist John Roebuck and his even more fruitful partnership with Matthew Boulton were instrumental in the development of the steam engine that literally powered the Industrial Revolution.

Other philosophical societies sprang up across the country. For example, the Manchester Literary and Philosophical Society was founded in 1781 to promote polite knowledge, rational entertainment, technological instruction, and professional occupation. The pursuit of scientific knowledge was seen as both transcendent (as a path to God beyond the material world) and as an intellectual ratifier of a new world order based on industrialism and the exploitation of the material world. Manchester was a new manufacturing city, and its elite citizens sought legitimation and an ideology of progress and utility that could represent their lives. The Manchester Lit and Phil (as it was called) was founded by medical men and included Quakers, Unitarians, and manufacturers. This was a society where provincial and mercantile concerns were more important than London directives. London scientists were not telling manufacturers how to run things; equally, manufacturers were not telling scientists what to do. Rather, these societies provided a controlled outlet for radical agitation in the first generation and a means of maintaining the power won in the second. Soon, Lit and Phils were founded in Bristol, the Potteries, Newcastle, and Edinburgh.

Classification Systems

One of the recurring concerns of those collecting artifacts for museums or fossils for geological theorizing was the question of how to categorize or classify their findings. Classification was a goal of the encyclopedists, matter theorists, and imperialists. Control of the world and its resources could come only from knowing the names of everything and where each type or individual fit into the larger scheme of things. Through the Middle Ages and Renaissance, classification had been based on the Great Chain of Being, which was a strict hierarchy descending from God, angels, people, animals, plants, and ending with the inanimate world. Scholars were no longer satisfied by this system and therefore sought new, more rational classification schema. These thinkers devised rational systems to classify first plants and animals and later chemical elements, participating in the great Enlightenment project to know everything.

Carolus Linnaeus (Karl von Linné) (1707–78), a Swedish botanist, was one of the most successful systematizers in the eighteenth century. He amassed a massive botanical collection (dried, rather than living, as opposed to the collection at Kew), receiving plants from other collectors all over the world. Linnaeus developed a classification system, first articulated in *Systema Naturae* (1735), based on increasing specificity: Kingdom, Class, Order, Genus, Species, Variety. In order to classify specific plants, he used "artificial classification"; that is, he based his system on attributes that were easily counted and measured but probably did not link the species in nature. All systematizers in the eighteenth century used artificial systems but hoped one day to find the natural basis for connections among species. As it turned out, it was only through adding time, as Darwin did, that such natural classification became possible. Linnaeus based his system on the sexual characteristics of plants, specifically by counting the number of stamens and pistils. This was not random choice, since sexual organs seemed fundamental in passing on certain basic characteristics. He then developed a binomial nomenclature of genus and species, which is still used today.

Linnaeus looked at the relationship of clearly related genera and species to express the concept of branching classification. That is, he argued that there was no clear linear progression or hierarchy from simplest to most complex organism, as the Great Chain theory

6.7 PAGE FROM LINNAEUS'S *SYSTEMA NATURAE* (1735)

posited, but that each genus, each species, was equally complex. He pictured the relationships of species laid out like countries on a map, with each species touching many others. At first, in *Philosophia Botanica* (1751), he argued for the fixity of species, seeing no transformation from one to another. By 1760 he had decided that a number of species might have common ancestors but believed that this was caused by hybridization only. Linnaeus saw classification as a closed system in which basically no new species could appear; therefore, all were theoretically knowable. He envisaged the map of nomenclature as a finite table, where systematizers could work to fill in the gaps, eventually producing full knowledge of all living beings. (This was very similar to the development in the nineteenth century of the chemical periodic table.) Thus, Linnaeus showed the natural affinity of species, based on similarity of characteristics, and mapped out an ambitious research program of naming all living things.

Although Linnaeus was first and foremost a botanist, he did transfer his nomenclature and his system to the naming of animals. Until then, animals had been categorized by the Aristotelian groupings of four-footed versus two-footed, plus birds and fishes. Linnaeus worked to find similar characteristics to link different species. Perhaps his most controversial move was the identification of one group of animals by the mammaries of the female half of the class (that is, mammals). This may have been influenced by the fact that Linnaeus was himself an influential member of the anti-wet-nursing campaign in Sweden, which encouraged upper- and middle-class women (like his own wife) to breastfeed their own children. By grouping together horses, dogs, apes, and humans into a single class, he set in motion a huge debate about the place of human beings in nature, both in God's plan and in the biological world.

The Study of Matter: Chemistry and the End of Alchemy

An interest in classification merged with issues of power over nature through understanding its inner forces in the study of matter. Contrary to the trend of mathematical physics, which had been organized and simplified by Newton into a series of axiomatic rules, chemistry lacked a central organizing conception or even a commonly accepted language and nomenclature; as a result, matter theory was in a state of chaos. More substances and processes were being discovered, but without a set of central organizing principles these simply meant there was more to be confused about. Industrial chemistry was still largely a craft or guild system of production, but the demand for materials as Europe began to industrialize pushed producers to expand production and look for new methods. Some materials – gunpowder, dyestuffs, acids, and shipbuilding materials such as pitch – were in such high demand that their production became concerns of national importance.

Alchemy, while still common, was under increasing attack by natural philosophers who sought to put the study of matter on a rational and experimental basis. Since Boyle's publication of *The Skeptical Chymist* (1661), these researchers had been working to establish methods for examining the material world that did not depend on hidden or occult forces. This was reinforced by the 1732 publication of the *Elementa chemiae* (*Elements of Chemistry*) by Herman Boerhaave (1668–1738). Translated into most major European languages, it was considered to be one of the foundational texts for chemistry until it was supplanted by the work of Lavoisier 50 years later. A further attempt to bring some order to chemistry was introduced by Pierre Joseph Macquer (1718–84). Trained as a physician and appointed to the Académie des Sciences in 1745, his work moved away from Boyle's approach and adopted a more Newtonian and corpuscular system. In 1751 he published two influential textbooks, *Elémens de chymie pratique* and *Elémens de chymie théorique*, and in 1766 produced the *Dictionnaire de chymie*. Although certain conceptions from Aristotle still lingered in the work of Macquer and his contemporaries, the emphasis was increasingly on experimental procedure and quantitative analysis.

Many of the leading chemists of the eighteenth century worked on pneumatic chemistry, studying "airs," or, as we call them, gases. There were a number of reasons for the primacy of the study of various airs. First, they were intimately linked to life, and the investigation of life was understandably popular. They were also linked to other processes – for example, combustion, rusting, and calcination – and were the by-product of many important operations such as brewing, smelting, and dye making. On an intellectual level, airs were the finest type of matter, so it was believed that understanding their structure and behavior would open the door to a more general understanding of matter. This followed the tradition of the corpuscularians, particularly Descartes, Boyle, and Newton, who had all argued that matter and even light (in the case of Newton) were made of extremely small particles.

Only one aspect of this growing list of substances was in any way unified, and that was the principle of combustion. The study of combustion (which included calx or rusting and respiration) had been investigated by a number of philosophers in the seventeenth and early eighteenth centuries; most had agreed that fire was a substance liberated from compound bodies when they were heated. In the older alchemical tradition, fire was considered an essence or spirit. In the later tradition of iatrochemistry, people such as Joachim Becher (1635–82) argued that it was a *terra pinguis*, or oil earth (somewhat akin to the Aristotelian fire element) that was combined with other matter to form the materials of the world. The more *terra pinguis* a material contained, the greater its potential combustibility.

Becher's earlier system was revamped by Georg Ernst Stahl (1660–1734), who had corresponded with him. In 1718 Stahl replaced *terra pinguis* with phlogiston, using the Greek root *phlogos*, meaning flame. He argued that metals were composed of calx and phlogiston, so that when the metal was heated, the phlogiston was released into the atmosphere and left behind the calx. All inflammable substances contained phlogiston, but phlogiston itself had properties that set it apart from other forms of matter. This system allowed chemical reactions to be charted.

Metal (*calx + phlogiston*) $\xrightarrow{\text{heat}}$ Calx + Phlogiston (*released to atmosphere*)

The phlogiston theory neatly explained many of the aspects of combustion. A substance would stop burning if it saturated the surrounding air with phlogiston (such air was then "phlogisticated"), as in the case of a candle in an enclosed jar. (See figure 6.8.) Combustion would also stop if all the phlogiston was expelled from the substance in the complete formation of a calx. Phlogisticated air would not support respiration, but the atmosphere did not become saturated because plants absorbed or fixed phlogiston from the atmosphere, returning it to the flammable wood. Supporters of the phlogiston theory found that substances were often heavier after combustion or calcination than before, suggesting that phlogiston had negative weight or positive lightness and was not attracted to the Earth like matter in the Newtonian system.

A number of researchers continued the work on airs throughout the eighteenth century. A technical development by Stephen Hales (1677–1761) made the collection of airs much easier. Hales introduced the pneumatic trough in which gases were collected by the displacement of water. While the method may have predated Hales, his use of it was quickly taken up by other experimentalists such as Joseph Black (1728–99) and Henry Cavendish (1731–1810). Black, working in Scotland, began his chemical investigations while researching his doctoral dissertation in medicine. He became interested in the relationship between acids and alkalis and their relation to "fixed air," or what is today called carbon dioxide. Black demonstrated that fixed air would not sustain combustion by pouring the

Phlogiston

 Element

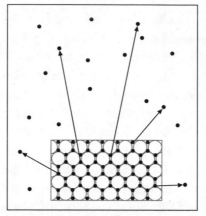

Dephlogisticated air supports combustion
by allowing phlogiston to move from the
area of high concentration in the element
to low concentration in the air.

Phlogisticated air is saturated,
so no combustion is possible.

6.8 PHLOGISTON

invisible gas onto a lit candle in a container, thereby putting out the flame. Using carefully measured experiments, he proved that fixed proportions of chemicals combined in chemical reactions and that fixed air was a component of atmospheric air, as well as being one of the gases produced when exhaling. By these demonstrations, he showed that atmospheric air was a mixture of gases and not elemental as it had been considered in the Aristotelian and many later systems.

Cavendish, after whom the great Cavendish Laboratory in Cambridge is named, identified the properties of "inflammable air" (now called hydrogen) in 1766 and distinguished it from a number of other gases that were known to burn. The properties of inflammable air suggested to a number of people that it was phlogiston. Around 1784 Cavendish was the first to demonstrate clearly that water was a compound, undoing another of the Aristotelian elements.

While the work of Black and Cavendish was wide-ranging, the greatest researcher on the nature of "airs" was Joseph Priestley (1733–1804), who isolated and studied more new gases than any other investigator. He had little formal training in natural philosophy, but he came into contact with a number of people who were interested in the subject such as Matthew Turner (d. 1788), a physician who had lectured on chemistry, and Benjamin Franklin. When Priestley took a position as librarian to Sir William Petty, second Earl of Shelburne, he had both a patron and a position that allowed him to work on his research. Between 1774 and

1786 he produced six volumes entitled *Experiments and Observations on the Different Kinds of Air*. He investigated what we would now call nitric oxide, hydrogen chloride, ammonia, sulfur dioxide, and oxygen, among others. He investigated the properties of phlogisticated air in water (seltzer or soda water) as a cure for scurvy. His most important work, as it would turn out, was on what he called "dephlogisticated air." When he heated a calx of mercury, he obtained what he thought was a new gas that was very combustible. He reasoned that this air contained so little phlogiston itself that the phlogiston in the combusted substance rushed to fill the void. The reverse of this, phlogisticated air, was so full of phlogiston that no more could enter, and thus combustion could not take place.

Priestley's chemical interests were closely tied to his radical political and religious views. He was a Unitarian, connected with the dissenter academy of Warrington Academy, later Manchester College, which offered science training as part of its goal of progress and rationality. Priestley believed in the idea of progress, the perfectibility of man, and the ability of humans to find out the truth about everything, starting with the natural world. He was, therefore, in favor of the abolition of repressive laws and was sympathetic to both the American and French Revolutions. Just as in the case of Hutton, Priestley became a more and more suspect character, especially once the British declared war on the French. On July 14, 1791, his house in Birmingham was vandalized by a mob (sometimes called the Priestley Riots). He left Britain in 1794 and settled in Pennsylvania, where he continued to support the phlogiston theory until his death.

While important work on airs was done in Britain, the center of this scientific research was in France, particularly Paris. By 1750 the Académie des Sciences in Paris had attracted some of the most influential scientists on the continent. Because there was a government salary for full positions and the number of places in the Académie were limited, there was much competition and political intrigue associated with membership. Into this world of salon society, scientific development, and reform movements came Antoine-Laurent Lavoisier (1743–94). He was a reformer, aware of the political implications of his work within science and for French society. He began his higher education in law, following in his father's footsteps, but was attracted to science, particularly chemistry. He gained an associate position in the Académie in 1768, and, although this position was the lowest rank and unpaid, he took it, deciding to devote his life to science. To finance his work, he took a share in the Ferme générale, a private tax farm, an organization that collected taxes for the government. Although his eventual elevation to Academician status provided him with a government salary, his personal fortune and his share in the Ferme made him independently wealthy. He eventually became a Farmer-General, one of the high officials in the organization. While this supported his research, his association with tax collection led to his death on the guillotine in 1794 when he was tried and convicted by a revolutionary court of being anti-revolutionary.

In 1771 Lavoisier married Marie Anne Pierrette Paulze (1758–1836), the daughter of one of the Ferme's partners. Marie was integral to Lavoisier's scientific and political life. She managed his affairs, learned English so that she could translate materials for him, and attended to laboratory work. She studied art with Louis David and was responsible for the engravings that accompanied Lavoisier's work. In keeping with salon culture, she hosted the twice-weekly meeting of intellectuals that Lavoisier entertained in their home. She continued this practice even after his execution. In this, she fits a pattern of women in science who work behind the scenes and who were often unnamed partners in the scientific enterprise.

Lavoisier, as a member of the Académie, was also a public servant, expected to place his intellectual abilities at the service of the state. Given his energy and interest in reform, he was more than willing to undertake reports on a wide variety of subjects, including a review of the water supply of Paris, the condition of prisons, adulteration of food, ballooning, and a range of industrial concerns such as the ceramic industry, glass making, and ink manufacturing. He also worked on the reform of agriculture, becoming a member of the Committee on Agriculture in 1785. In connection with his work for the state and his prowess as a chemist, he was made director of the Royal Gunpowder and Saltpeter Administration and was set the task of improving French gunpowder, which was often of very poor quality and whose raw materials were difficult to obtain. In this position he was given a fine house at the Paris Arsenal and ample space for a laboratory.

When Priestley visited Paris in 1774, he discussed his work on dephlogisticated air with Lavoisier, who had been looking for the component part of the atmosphere that supported respiration and combustion. Ironically, Lavoisier used Priestley's discovery to destroy the phlogiston theory. By 1777 he had concluded that "eminently respirable air" was converted to "fixed air" by combustion and respiration. Because of its relation to acids, he named this air *oxygène* ("acid former" in Greek). By 1778 he demonstrated that atmospheric air was a combination of this respirable air and an inert air. This opened the door for a number of further experiments, including a demonstration that water was composed of oxygen and Cavendish's inflammable air. These experiments convinced Lavoisier that the phlogiston system could not work, and in 1783 he submitted a paper to the Académie entitled *Reflections on Phlogiston* that set out the problems of the old theory and how his oxygen system solved them. This controversial position gained Lavoisier not only a number of supporters, including Joseph Black, but also stiff resistance from prominent chemists, including Macquer and Priestley himself.

Discarding phlogiston left a problem concerning heat. If calx and respiration were simply a chemical combination, what was heat? With the mathematician/physicist Pierre-Simon Laplace, Lavoisier reformulated the place of heat in the system, introducing the concept of caloric to replace the principle of phlogiston. To quantify the production of heat, they created the ice calorimeter, which used the latent heat of fusion to demonstrate the expiration of heat from chemical reactions and respiration. (See figure 6.9.) Latent heat had been identified by Joseph Black in 1760 when he noticed that at the melting point additional heat melted ice without raising its temperature (until all the ice was melted when the water began to heat up). Thus, a specific quantity of heat was needed to melt a certain volume of ice into water. By measuring the amount of water, the amount of heat could be calculated. Lavoisier considered caloric to be an imponderable fluid that caused other substances to expand when it was added to them. For example, during the combustion of carbon and oxygen, the resulting carbon dioxide had the combined weight of the original substances and gave up the caloric as heat and light. Although the concept later proved to be wrong, Laplace and Lavoisier's measurement of energy is the basis for the calorie of food energy.

Lavoisier and a number of his supporters decided that to put chemistry on a rational and useful foundation – in this case meaning his system and the rejection of the phlogiston theory – they would have to reform all of chemistry. In keeping with humanist and Enlightenment philosophy, they started with language, and in 1787 in collaboration with Claude Louis Berthollet (1748–1822), Antoine François de Fourcroy (1755–1809), and L.B. Guyton de Morveau

6.9 LAVOISIER AND LAPLACE'S ICE CALORIMETER FROM LAVOISIER'S *ELEMENTS OF CHEMISTRY* (1789)

(1737–1816), Lavoisier published *Méthode de nomenclature chimique*. This work attempted to unify and systematize the naming of chemicals and elements, replacing old common names with Latin and Greek roots. Thus "vitriolated tartar," "sal de duobus," and "arcanum duplicatam" all became potash. In addition to the root names, compounds were distinguished with various suffixes to indicate their classes. Salts formed by sulphuric acid were called sulphates, while those formed from sulphorous acid were sulphites. Using the system of nomenclature in effect meant accepting Lavoisier's underlying oxygen theory, thereby ensuring that his system would be the new path for chemical research.

Initially, opposition to Lavoisier's new system was very strong, especially among older chemists. His campaign of persuasion was carried out not only in the formal world of publishing and the Académie des Sciences but also in the salons. In the twice-weekly meetings hosted by Marie he entertained virtually every important natural philosopher who visited Paris. As director of the Académie from 1785, he manipulated the organization of the chemistry section so that it was made up only of anti-phlogistonists. When the editorial control of the *Journal de physique* was taken over by phlogistonists in 1789, Lavoisier and his disciple Pierre Adet (1763–1834) founded the *Annales de chimie* in an effort to improve both the reporting on and quality of chemistry. This journal continues to be a leading scientific publication today.

In 1789 Lavoisier published his most influential book, *Traité élémentaire de chimie* (*Elements of Chemistry*). This brought together all aspects of his work, introducing his nomenclature, his experimental system and apparatus, and his methods and standards of measurement, and including an extensive compilation of all elements and compounds recognized under his system. Widely read and quickly translated, it gave the death blow to the phlogiston theory. Essentially, no young chemist or person interested in matter theory could claim to be current in the field without being acquainted with Lavoisier's system. His nomenclature was so functional that it quickly came to be the most widely used system, carrying his theory of chemistry with it. Equally, measurement and experimentation became integral to chemical research, paving the way for very different projects than were possible with the older qualitative practice.

The first stage of the chemical revolution ended with the French Revolution. Lavoisier, who was a reformer and political moderate, had hoped to use science to support a new and more progressive France. He had worked hard to make France powerful, both intellectually and in practical terms with his system for producing gunpowder, agricultural reform, and geological work. When the Revolution degenerated into the Terror, Lavoisier's link to the old regime through the tax farm made him suspect in the eyes of the radicals. He was denounced and arrested along with 27 other members of the Ferme. Their trial lasted just a few hours, and, although the charges were without substance, the convictions were really a foregone conclusion, and nothing Lavoisier or his friends said could sway the outcome. When Lavoisier went to the guillotine on May 8, 1794, France and science lost one of its most powerful minds. Lagrange said of Lavoisier's death that "it took only a moment to cause this head to fall and a hundred years will not suffice to produce its like."[3]

With the rise of chemistry as a public and systematic study, inquiry into the transmutation of materials faded from the realm of serious consideration by natural philosophers and the increasing number of academicians studying science. One final episode effectively closed the door on alchemy and alchemists. This was the affair of James Price, born James Higginbotham in London in 1752. Although he graduated from Magdalen Hall, Oxford, as a medical doctor in 1782, he had been elected a Fellow of the Royal Society a year earlier, in part because of his work on chemistry. After his graduation, he invited some important people to his home, claiming to have transmuted mercury to gold. He added a mysterious white powder to 50 times its weight in mercury, mixed in some borax and niter, and heated the mixture in a crucible. It produced an ingot of silver. When Price followed the same procedure with 60 times the weight of mercury, the product was gold. The ingots were found to be genuine metal and shown to the king.

This created a sensation, and Price introduced a new twist to the old con of alchemy. Rather than gaining patronage directly for his work, he wrote a pamphlet on his experiments that became a bestseller. By 1783 Price said that his supply of white powder was exhausted, and the cost of making more would be too great in expense, labor, and drain on his health, since he hinted that some spiritual effort was required to make it.

Joseph Black reviewed Price's work and said that it was a mass of errors. He was astonished, he said, that Price had ever received a medical degree. Supporters and detractors rushed into the fray, until a committee was struck to investigate the situation. Price refused for some time to cooperate, but pressure from the Society and his friends forced him to accept the investigation. He had six weeks to prepare, but it was all for nothing. When the team of three observers from the Royal

3. J.J. O'Connor and E.F. Robertson, "Joseph-Louis Lagrange," *MacTutor History of Mathematics*, last modified January 1999, https://mathshistory.st-andrews.ac.uk/Biographies/Lagrange/.

Society arrived, he showed them into his laboratory, excused himself, and left. He drank a vial of hydrocyanic acid, returned to the laboratory, and dropped dead at the feet of the investigators. In effect, alchemy died with him, since after this incident no scientific society in Europe would notice alchemical claims except to debunk them.

Conclusion

The French Revolution and the Europe-wide warfare that followed put an end to the more radical or "free-thinking" positions of a number of natural philosophers. Rational secular explanations for natural phenomena, planetary development, and the evolution of species became dangerous ideas in the 1790s. Nonetheless, natural philosophy had transformed in this Enlightenment period. Quantitative analysis, logical and rational classification systems, measuring and explaining phenomena such as electricity and heat that had once been seen as esoteric – all this was the legacy of a century of development.

One of the lasting reforms that eventually transcended the chaos of revolution and warfare was the metric system of measurement. Even within a single country, there was often no uniform set of measurements. France, for example, had around 300 different units of weight. As early as 1670 Gabriel Mouton, a French vicar, proposed to reform the systems of weights and measures using scientific principles. In 1742 Anders Celsius (1701–44), a Swedish astronomer, introduced the centigrade thermometric scale, the first important decimal (base ten) measuring system to be widely adopted in science. In the reform-minded era of the 1790s, Thomas Jefferson proposed a decimal-based system for measurement in the United States, and while this idea was not enacted, the American mint introduced decimal currency in 1792. In 1790 Louis XVI of France authorized an investigation into the reform of French weights and measures by the Académie des Sciences. Five years later the French Republican government officially adopted the metric system. Although the measurement system of science we know today as SI (*Système internationale d'unités*) was the product of international conferences in the nineteenth century, the concept of a measuring system that was both uniform in units and integrated (a fixed volume of water equaling a specific mass, for example) was founded on the revolutionary reforms of the Enlightenment.

Just as the Enlightenment and the Industrial Revolution created whole new categories of jobs, so too in this age we see the creation of scientists: people who by profession, education, and association made the new science their life's work. Scientists concentrated on the utility of their knowledge about nature, investigating ways in which their study could improve the lives of citizens, the power of the state, and the bottom line of joint-stock companies. Science was now inexorably intertwined with the great imperial project of the nineteenth century.

Essay Questions

1. How did the ideas of the scientific revolution influence the Enlightenment?

2. How did geologists argue the Earth had been created? What was their evidence?

3. How did scientific collecting change the development of science?

4. What did Lavoisier revolutionize in chemistry?

CHAPTER TIMELINE

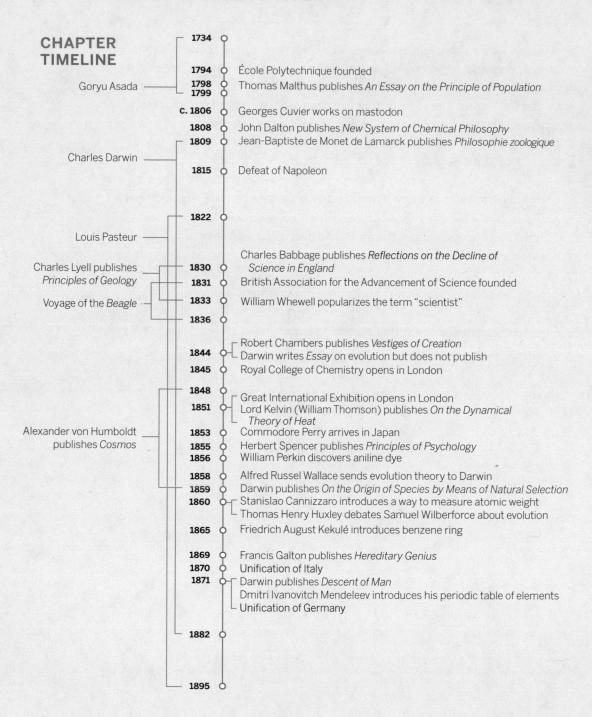

	1734	
	1794	École Polytechnique founded
Goryu Asada	**1798**	Thomas Malthus publishes *An Essay on the Principle of Population*
	1799	
	c. 1806	Georges Cuvier works on mastodon
	1808	John Dalton publishes *New System of Chemical Philosophy*
	1809	Jean-Baptiste de Monet de Lamarck publishes *Philosophie zoologique*
Charles Darwin		
	1815	Defeat of Napoleon
	1822	
Louis Pasteur		
Charles Lyell publishes *Principles of Geology*	**1830**	Charles Babbage publishes *Reflections on the Decline of Science in England*
	1831	British Association for the Advancement of Science founded
Voyage of the *Beagle*	**1833**	William Whewell popularizes the term "scientist"
	1836	
	1844	Robert Chambers publishes *Vestiges of Creation*
		Darwin writes *Essay* on evolution but does not publish
	1845	Royal College of Chemistry opens in London
	1848	
	1851	Great International Exhibition opens in London
		Lord Kelvin (William Thomson) publishes *On the Dynamical Theory of Heat*
Alexander von Humboldt publishes *Cosmos*	**1853**	Commodore Perry arrives in Japan
	1855	Herbert Spencer publishes *Principles of Psychology*
	1856	William Perkin discovers aniline dye
	1858	Alfred Russel Wallace sends evolution theory to Darwin
	1859	Darwin publishes *On the Origin of Species by Means of Natural Selection*
	1860	Stanislao Cannizzaro introduces a way to measure atomic weight
		Thomas Henry Huxley debates Samuel Wilberforce about evolution
	1865	Friedrich August Kekulé introduces benzene ring
	1869	Francis Galton publishes *Hereditary Genius*
	1870	Unification of Italy
	1871	Darwin publishes *Descent of Man*
		Dmitri Ivanovitch Mendeleev introduces his periodic table of elements
		Unification of Germany
	1882	
	1895	

SCIENCE AND EMPIRE

7

The nineteenth century was the great age of European empires. Although European exploration and colonization started much earlier, European nations, especially those on the Atlantic coast, now dominated all parts of the globe economically, militarily, and politically. The steam engine, the telegraph, and the factory conquered time, space, and material desires. No part of the globe was beyond the reach of empire, and the flags of those empires were carried to the most remote and challenging places. As Western Europe underwent rapid industrialization, it turned more and more to colonial holdings for natural resources and captive markets. Both industrialization and colonialism helped to spur the development of science, which offered the ability to know the world better and revealed ways to turn the natural resources of the colonial holdings into wealth. This was not strictly a one-way exchange, as non-European countries started to adopt ideas and practices, including the study of nature, from their contacts with European thought. At the same time, for those European nations that had limited colonial holdings, science offered a way to deal with economic disadvantage by creating new tools and techniques to solve problems that resulted from a lack of cheap natural resources.

The big winner of the colonial game was Britain. After the defeat of Napoleon in 1815 the strongest rival to British power was subdued, largely leaving the world open to unfettered British exploitation. France had been defeated not just on the battlefield but also in the shops and factories. British industry, backed by its colonial strength, out-produced France, so that every bullet, tent, and naval vessel cost less and was produced faster and in greater numbers than France could match. Although the American colonies had broken away from British rule in 1776, they represented only a minor part of British holdings when compared to its control of India and parts of Africa, the Middle East, and Asia. And, despite a century of suspicion and conflict between Britain and the United States, heightened during the War of 1812 and British trade with the Confederacy during the American Civil War, Britain remained the United States' largest trading partner throughout the era.

Paradoxically, Britain's power helped pave the way for a series of political realignments on the continent that came to challenge British supremacy. The Franco-Prussian War of 1870–1 led to the unification of the German states by Otto von Bismarck, while the Italian states were unified by Garibaldi and came under the political control of Victor Emmanuel II by 1870. The wars of the twentieth century were rooted in the power blocks created in the nineteenth century.

As important as the "great game" of empire was, equally remarkable was the fact that those who stayed back home were reading reports and seeing images of those places within days or even hours of the events. In 1804 Richard Trevithick (1771–1833) and Matthew Murray (1765–1826) built the first steam engine tramway locomotives, starting a transportation revolution that saw hundreds of thousands of kilometers of rail laid by the end of the century. The first transatlantic voyage by a steam-powered ship was made by the *Savannah* in 1819, and by the middle of the century steamships were replacing sailing ships. As swift as the locomotives and the steamships became, they could not beat the speed of the telegraph for moving information around the globe. Between rail, steamship, and telegraph, the information flow needed to control global empires and international trade was now possible.

Collecting and Classifying: Biology and Empire

With the major expansion of European imperialism, both scientists and the general public became more and more interested in collecting the many new and exotic species discovered. The mania for the bizarre and unique continued, and curio cabinets and displays of exotic birds, bugs, and hunting trophies were a feature of many middle-class homes. For the serious researcher, collecting gave way to a search for order and connection. After Linnaeus, the collecting sciences of the eighteenth century became the biological sciences of the nineteenth. This was deeply connected to the notion that classification and understanding were part of the process of controlling and exploiting and, so, were bound up with the imperial project.

We can trace the transformation of the study of the living world from eighteenth-century natural history to nineteenth-century biology through the career of one man, Alexander von Humboldt (1769–1859). The son of a Prussian military officer, Humboldt chose the career of a scientific traveler and naturalist rather than the diplomatic life his father had envisaged. He was influenced by the naturalists' voyages with Captain Cook and, in 1799, set out on his own prolonged travels through Spanish America. The result was a series of travel and natural history books that were bestsellers. But Humboldt's descriptions of flora and fauna differed markedly from those of Banks or other eighteenth-century naturalists. He combined exact measurement using precise instruments with intensive fieldwork and an overall concept of the interconnectedness of all living things. His greatest articulation of this came in his final multivolume work *Cosmos* (1848–59), which was very influential for future scientists. Humboldtian science, as some historians have called it, was instrumental in engendering a strong reliance on fieldwork among biologists, especially in the United States, and in creating a proto-ecological understanding of the living world.

The categorization of new flora and fauna continued with the discovery of additional, and potentially exploitable, species. Naturalists and explorers searched out and found uses for these new plants, much as Hans Sloane had done with chocolate. Often this involved interacting with Indigenous peoples and appropriating their knowledge of their natural world. In many cases this interaction began as an exchange of knowledge among equals and gradually moved to a deeply hierarchical and exploitative relationship. One of the most important discoveries was that of quinine, the only known treatment for malaria; in an era of greater European exposure to the disease, it helped make imperialism in the equatorial regions possible. Quinine was itself a product of colonialism, since it came from the bark of a tree found in Peru and was introduced to the Spanish by the local people in the 1630s. The botanical name given to the plant by Linnaeus was *Cinchona*, after Countess Ana of Chinchon, the wife of a Viceroy of Peru in 1638. Although the story is probably apocryphal, Countess Chinchon supposedly became ill with malaria and was successfully treated with quinine bark, thus ensuring the spread of its use. Because *Cinchona* was hard to grow and expensive, the efforts to produce artificial quinine became one of the great quests of organic chemistry.

As the empires encountered, traded with, and on many occasions subjugated other peoples, the scientific move to classify came to include human beings as well.

7.1 BLUMENBACH'S RACE DISTRIBUTION

Linnaeus had paved the way for this through his classification of humans as thinking mammals (*Homo sapiens*). Further, he claimed that there were four races of humans: Europeans, Asians, and Africans (in keeping with the Greek tripartite division of the world), plus the New World race, the Americans. Johann Friedrich Blumenbach (1752–1840), one of his protégés, furthered this categorization by developing a model of relations among these peoples. He believed that all people were of one species, but that there was a hierarchy of perfection among the races. Blumenbach used the people from the Caucus mountains, the most beautiful people he had ever seen, as the race with maximum perfection and then plotted the other races as degenerating from the norm as they moved away to the east and west. In order to have a symmetrical model, he added a fifth race, the Malays. (See figure 7.1.)

Although Blumenbach had not intended his categorization to indicate a hierarchy of development, or to imply that somehow the races below the Caucasians were closer to animals, nineteenth-century racial theorists soon took this pyramid as indicative of biological and cultural achievement and as an excuse and opportunity for exploitation. Perhaps the most egregious example of this is the story of Saartjie (Sarah) Baartmann, a Khoi native of South Africa, who was taken to Britain in 1810 by William Dunlop, a British naval surgeon, to be displayed in sideshows as the "Hottentot Venus." This resulted in a scandal, since slavery had been outlawed in Britain, and eventually Sarah's show moved to France. Scientists such as Cuvier were very keen to examine her, especially because of her unique anatomical features, but Baartmann refused. She eventually died alone as an impoverished prostitute in the streets of Paris in 1816. After her death, she was dissected by a French surgeon, and various of her body parts were on display in the Natural History Museum in Paris until 1949 and then at the Museum of Mankind in Paris until 1974.[1]

----

1. Baartmann's remains were taken off display and repatriated in 2002. She was buried near the Gamtoos River in Hankey in the Eastern Cape, South Africa. Her return to Africa was seen as both an indictment of the racism of the colonial era and the end of a terrible episode in human exploitation.

Catastrophe or Uniformity: The Geological Record

Although the study of people, plants, and animals from far-off locations attracted a great deal of popular and scientific attention, one of the most important areas of research developing in Europe was geology. It flourished under imperial competition because of its link to industrial development associated with coal, iron, and other exploitable mineral resources. Britain, for example, embarked on major field projects to understand, map, and name the strata, and imperial geological surveys were undertaken in Canada, Australia, New Zealand, and India.

Just as they had in Enlightenment Europe, geologists continued to delve into the history of geological change. The debate over causes – the Neptunism and Vulcanism of the eighteenth century – was transformed into a more quantitative discussion of the rate and type of change. The two schools of thought – Catastrophism and Uniformitarianism – that emerged from this debate relied on the idea of geological formations created by either sudden or gradual change. The crux of the problem was this: had the Earth once been hotter, had volcanic activity once been greater, or had the Earth always operated in the way it now did? Hutton had said that forces now in operation explained all geological change, but he had been labeled a dangerous radical by the end of the century, and most early nineteenth-century geologists disagreed with him. Not until Charles Lyell's work in the 1830s did someone again bring forward the idea that forces currently in operation were responsible for changes in the Earth's configuration historically.

Georges Cuvier (1769–1832) was the most prominent of the catastrophists. He was a professor of anatomy at the newly created National Museum of Natural History in Paris. As a state employee, through his position and work he demonstrated the prominence of France in imperial competition, even after Napoleon was defeated. In addition to his geological ideas Cuvier studied comparative anatomy, establishing that the parts of individual animals must work together and, therefore, that every conceivable permutation or combination of animal parts was not possible. For example, animals with carnivorous teeth had to have suitable stomachs for digesting meat. He became famous for his ability to reconstruct an entire animal from a relatively small part of its skeleton.

When Cuvier was sent some odd fossils from Paraguay, he determined that these most closely resembled the modern sloth, although the remains suggested a giant and now extinct species. He quickly realized that the only way to prove extinction was by using this sort of giant remains, since earlier debates concerning marine fossils always remained inconclusive, given the possibility that the sea

Grand MASTODONTE. PL. V.

7.2 CUVIER'S MASTODON FROM 1806

creatures might still be living in unexplored ocean depths. Cuvier looked at various elephant-like remains and, using comparative anatomy, established first that African and Indian elephants were different species and, second, that the remains found near the Ohio River in the United States indicated a different species altogether. He called this a mastodon, while he named a different but related specimen from Siberia a mammoth. (See figure 7.2.) Eventually, frozen carcasses of this Siberian species were found, complete with woolly coats, indicating that they were not tropical animals who had strayed far from their natural habitat, since they were clearly creatures from a cold climate.

So, Cuvier asked, why had they died out? Most of the remains of giant mammals found near Paris, particularly ancient hippos and rhinos, were in gravel pits. Therefore, he reasoned that there must have been a sudden "revolution," probably a flood, that had killed them. He and his fellow researchers discovered that the local gypsum quarries showed evidence of seven progressive floods, with alternations between freshwater and saltwater fossils, and from this determined that there must have been a series of catastrophes. Cuvier used recent work on the Alps, which demonstrated through an analysis of the strata column that they must be of recent origin, to suggest that their emergence might have been a cause of one of the catastrophes. He stressed, however, that these catastrophes were local, not universal, and therefore were not tied to the biblical flood. Although he could not identify the causes of all the revolutions, he was convinced that they were ultimately knowable.

Cuvier stressed the progressive aspect of the history of life on the planet, with extinction as part of the package. As more fossils were discovered, by Cuvier in France and William Smith (1769–1839) in Britain, they discovered that in the gypsum layer under the gravel pits were fossil mammals even more different from present animals than the hippos and mastodons had been. For example, Cuvier found the remains of an animal that looked like a combination of a tapir, pig, and rhino. The remains in the Tertiary rocks were largely mammals, while the

Secondary rocks contained mostly lizards. In fact, Cuvier found a number of lizards occupying diverse parts of the ecological scene: a flying one, which he named "ptero-dactyle"; a swimming one, which he called "ichthyosaurus"; and a walking one, which he named "iguanodon." And in the Primary rocks, nothing. Here was a clear progression from one sort of life on Earth to the next.

Some geologists continued to equate the catastrophe evident at the end of the Pleistocene Age with the biblical flood, but most followed Cuvier's lead in ignoring this religious context. They concentrated instead on the evidence that life-forms had progressed through different stages, and it seemed more and more likely that the forces responsible for the mass extinctions must have been much more powerful and of a different kind than were evident today. This was confirmed by physicists who were looking at theories of heat radiation and who began to argue that the Earth had originally been much hotter and was gradually cooling down.

7.3 1846 RECONSTRUCTION OF AN IRISH ELK
This extinct deer, sometimes called an elk, was over three meters tall.

English geologist Charles Lyell (1797–1875) disagreed with this idea that the past was of a different kind than the present and therefore unknowable. He sought to maintain the rationality of science by developing a theory that was totally consistent and that enabled scientists to understand nature through observation in the present. In *Principles of Geology* (1830–3), a three-volume work whose title consciously referred back to Newton's *Principia*, Lyell expounded his theory of uniformitarianism. He argued that gradual cumulative geological change could account for extinction and the progress of species in the fossil record. He used his own research on Mount Etna in Sicily and the work of George Poulett Scrope (1797–1876) on the French mountains and the gradual creation of valleys through lava flow to show that gradual geological change corresponded to the gradual extinction of species in the same area. He also argued that the discovery of large mammals that had escaped extinction at the last supposed catastrophe, such as the giant Irish "elk" found in peat bogs above the Pleistocene level, showed that catastrophes were not the only cause of extinction and, perhaps, had not occurred. (See figure 7.3.)

Lyell's theory had three separate facets. The first, *actualism*, stated that forces of the kind now in action had created the world as we see it. This was reminiscent of Newton's claim for the universality of forces now in action to explain the structure of the universe. The second, *uniformitarianism*, established that forces of the same degree as today were at work in the past. That is, Lyell refuted the claim that things had been hotter or more violent in the past. The result of these two axioms was the third facet: that the world was in a *steady state*. Just as Hutton had said earlier, Lyell claimed that there was no progress, no direction, to the changes in the world. Unlike Hutton, however, Lyell had a much richer set of paleontological data, and therefore his was a much more difficult stance to support than it had been for his predecessor. Lyell was forced to stress the imperfection of the geological record and to claim that dinosaurs might be found at some other level in future investigations.

While the actualist and uniformitarian aspects of Lyell's theory were very attractive to geologists and biologists, the non-progressionist stance was a hurdle few could overcome. Lyell had to argue for the stability of species; that is, that there had been no evolution. But the work of many scientists, from Cuvier on, seemed irrefutable. As well, Lyell seemed to have brought back the idea of a "system," an idea that had been discredited by the work of people like Humboldt in favor of fieldwork.

Despite problems with his approach, Lyell's work was widely influential because of the attraction of his methodology of comparing the past with the present. The argument that this methodology was sufficient to understand nature guaranteed the autonomy of science from philosophy or religion. More significantly, Lyell was already an influential member of the scientific community. He was well known in the British scientific establishment, as a member and eventually president of the Geological Society of London, as a fellow and Royal Medal winner of the Royal Society, and finally as president of the British Association for the Advancement of Science (BAAS). Politics also favored Lyell's views. Uniformitarianism could be seen as bolstering conservative politics, keeping working classes in their place, as opposed to catastrophism, which fit more with the French view of the possibility of social and political improvement through revolution. Therefore, Lyell's theory did not disturb middle-class gentlemen, like Charles Darwin, who made up the scientific establishment. By contrast, on the continent scientists continued to follow Cuvier's theory, especially since it had the backing of physicists.

While these large theoretical debates raged, most nineteenth-century geologists spent their time on slow, steady fieldwork, sorting out the stratigraphy of the geological column. Field geologists, such as Roderick I. Murchison (1792–1871) and Adam Sedgwick (1785–1873) in Britain, spent many summers tramping through the countryside looking for exposed strata. This work turned out to be complex, since specific layers could overlay at one place, underlay at another, and be completely absent at a third. The competition to classify the various geological layers became intertwined with issues of political and professional power, especially seen in competing nomenclatures. In the end the strata were named for the British counties in which they were found (Cambrian for Wales, Silurian for the British tribe who had fought the Romans near the Welsh border, and Devonian for Devon). Murchison announced his identification of the Silurian system at the inaugural meeting of the BAAS in 1831.

The Question of the Origin of Species

One of the most perplexing questions arising from the new earth sciences and spurred by imperial expansion and exploitation was that of species. Where did new species come from? They seemed to appear without prior warning in the fossil record and to disappear just as rapidly. Was there a series of new creations? Did those species still exist somewhere undetected in the world? (This seemed unlikely in the case of dinosaurs or large mammals.) And what were species in any case? One of the earliest attempts to answer some of these questions came from a French scientist, Jean-Baptiste de Monet de Lamarck (1744–1829), who worked at the Jardin du Roi, as did Buffon and later Cuvier. Lamarck denied the possibility of extinction, arguing instead that one species transformed into another through evolution. In 1809 he published *Philosophie zoologique*, in which he articulated the evolutionary theory that has come to be called Lamarckianism. For Lamarck, the environment influenced the development of various characteristics and, thus, was of prime importance in evolutionary change. When the environment changed, forces internal to the individual plant or animal encouraged physical changes to take place to ensure adaptation to the new conditions. For example, if a short-necked ancestor to a giraffe came to live in an environment where all available food grew close to the top of very high trees, the internal forces within that giraffe would encourage the growth of its neck over time. Most crucial for Lamarck's theory, of course, was that these changes acquired in one generation could be inherited by the next; this was usually called the inheritance of acquired characteristics. Thus, the offspring of this stretching giraffe would be born with a slightly longer neck, the next generation with an even longer neck, and gradually the present-day giraffe would result. There was no extinction, since long-gone forms simply evolved into something else. Lamarck in this way

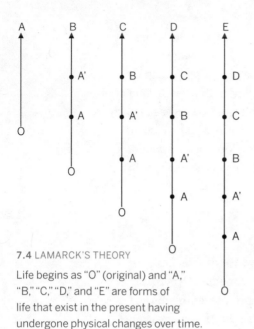

7.4 LAMARCK'S THEORY

Life begins as "O" (original) and "A,"
"B," "C," "D," and "E" are forms of
life that exist in the present having
undergone physical changes over time.

resurrected the Great Chain of Being, without its religious connotations, since he argued that every species began at the most primitive, and the existence of so many different species today was accounted for by a series of spontaneous generations of the most primitive forms. (See figure 7.4.)

Lamarck was reviled for this theory, especially by his enemy Cuvier. While Lamarck's theory serves to illustrate the fact that people in the early nineteenth century were thinking about evolution, he had little immediate influence on the field. For example, Charles Darwin owed little to his ideas. However, the theory made its mark when many prominent American biologists in the early twentieth century adopted it and it became known as neo-Lamarckianism.

In a similar way, both Erasmus Darwin (1731–1802), Charles's grandfather and member of the Lunar Society, and Robert Chambers (1802–71), who espoused evolutionary theories in early nineteenth-century Britain, demonstrate the prevalence of evolutionary thinking rather than serving as some linear path to Darwinian theory. Erasmus Darwin's ideas, published in a long poem, *Zoonomia* (1794–6), emerged from the radical thinking of the late Enlightenment and got him into trouble as a dangerous Jacobin free-thinker. Chambers's *Vestiges of Creation*, published anonymously in 1844, and based on a Laplacean view of an expanding universe, was wildly popular among the general public but ridiculed by scientists. Chambers was seen as an outsider, with no proper professional standing, and thus with no right to make such speculative claims. Darwin took this lesson to heart and worked to establish his credentials before publishing any "wild" theorizing.

Darwin and Evolution by Natural Selection

Charles Darwin (1809–82) was born into the ranks of the scientific elite, both socially and intellectually. He was the son of a rich doctor, the paternal grandson of an important member of the Lunar Society, and the maternal grandson of Josiah Wedgwood, the great British industrialist. Later, he married his cousin,

7.5 DARWIN'S VOYAGE ON THE HMS *BEAGLE* (1831–1836)

Emma Wedgwood, thereby solidifying his tie to the British manufacturing elite. He first attended the University of Edinburgh studying medicine but found surgery (in an era before anesthesia was common) too disturbing. He moved to Cambridge with the idea he would become a minister in the Church of England. An indifferent student, he first developed his enthusiasm for biology, and his life's work, when he spent one summer on a geology tour of Britain with his professor, Adam Sedgwick. He turned with alacrity to natural history and worked with botanist John Stevens Henslow (1796–1861), establishing his own collection of beetles. In 1831 Henslow recommended this eager and gentlemanly student to Captain FitzRoy as a naturalist and companion on the voyage of the HMS Beagle. Darwin would never have been recommended, or accepted, had he not belonged to the correct gentlemanly class.

The voyage of the Beagle (1831–6) changed Darwin's life. As he boarded the ship, Henslow handed him the first volume of Charles Lyell's *Principles of Geology*, a book that convinced him that forces in action today were responsible for all the changes of the past. He accepted uniformitarianism and actualism, although he never believed in Lyell's steady-state hypothesis. Darwin always saw natural change as directional. When he experienced an earthquake in Concepción, Chile, he was convinced that forces in action today could be very powerful and disruptive. He encountered a plethora of new and beautiful species, including fossils of the giant armadillo; on the Galapagos Islands, off the coast of Ecuador, he noticed that the species of tortoises and finches differed from island to island (although he did not really understand their classification or significance until later, in museum settings in London).

Darwin returned full of new ideas and soon established himself as part of the scientific profession through his published papers, first in geology and later in biology. His first scientific paper presented his theory of coral-reef formation, which followed Lyell's uniformitarianism and claimed that coral reefs were formed as an oceanic island submerged. Since most coral can only live close to the surface of the ocean, it would constantly grow on top of the submerging rock in layers to form a reef. This elegant explanation established Darwin's credentials with the Royal Geological Society.

Darwin began to keep a series of notebooks in which he puzzled over the relationship of species to each other – in fact, the whole question of just what a species was. In 1838 he read *An Essay on the Principle of Population* (1798) by Thomas Malthus, and the mechanism for evolution suddenly appeared clear. Malthus had argued that food supply would at best increase arithmetically (1, 2, 3, 4, 5 …) but that the population would increase geometrically (1, 2, 4, 8, 16 …) with the ultimate conclusion a life-and-death competition for resources. While Malthus had developed this theory in order to explain the population crisis he saw looming in Britain, Darwin immediately saw its application to the plant and animal world.

Darwin's theory, worked out between 1842 and 1844, is often called the theory of evolution by natural selection. He began from the premise that evolution had indeed taken place; his trip to South America had convinced him of this. So how had it happened? First, he argued, variation existed within a population of individuals. You can see this by observing domestic animals, such as the pigeons he bred. These variations were random, continuous, and small. Second, certain variations turned out to be advantageous in particular environments, and nature selected those variations – this is called natural selection. Since Malthus had shown that only a tiny fraction of the number of individuals born survive, there had to be some reason why some survived and others did not. Somehow, their variation equipped them better for survival than others in the population. For example, some birds were born with sharper beaks, which allowed them to burrow for insects; this was an adaptive characteristic in an environment where there were many burrowing insects and other food was scarce. This variation was passed on to the next generation, until more and more birds had this characteristic sharp beak and blunt-beaked birds were replaced or became extinct.

Darwin described an intraspecies struggle in which each individual in a specific species competed with other individuals in that same species for scarce resources. Other potentially dangerous species simply constituted the environment in which variation occurred. Nature was "red in tooth and claw," as Alfred, Lord Tennyson, put it,[2] because each day presented a competitive struggle for existence in which winning was everything. There was no second chance, no going back, because once you were extinct, you were gone forever. The result was branching evolution, where a common ancestor might give rise to numerous different species as different variations filled different ecological niches. (See figure 7.6.) Darwin answered the question of the definition of species by including time, since the map of species, as envisaged by Linnaeus, was now a snapshot in time of a continuing process, and the connections between closely related species, genera, and so on were due to their earlier common ancestry.

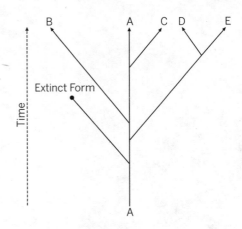

7.6 DARWIN'S SYSTEM

Darwin's system starts with an initial form that then produces related variants over time. Some variants may go extinct and are lost from the system.

This theory of struggle and competition corresponded very closely to the capitalist and imperialist struggle for resources seen in the clash of nations taking place around Darwin. He was influenced by earlier discussions of evolution and geology, but he was also a member of a rich mercantilist family, living in a nation prospering from industrialism and in major expansionist competition with other European nations. His theory was a product of a particular time and place, as much as it was the inspiration of one scientist.

Although Darwin wrote his theory of evolution by 1844 in a 230-page "*Essay*," he took no steps to publish. He may have feared the kind of ridicule that Chambers endured. He was also struggling to establish his credentials as a scientist among his peers in Britain. He spent the intervening years conducting a massive research project on barnacles, which, with his earlier coral-reef theory, garnered him much respect among his fellow scientists. In 1858, however, another naturalist, Alfred

2. Alfred, Lord Tennyson, "In Memoriam," in *Alfred, Lord Tennyson: An Authoritative Text, Background and Sources of Criticism*, ed. R.H. Ross (New York: Norton, 1973), 36.

CONNECTIONS

Science and Class: Wallace and Collecting

Alfred Russel Wallace came from a very different family background than Charles Darwin. He was not able to pursue a scientific career as a gentleman amateur, but rather earned his living collecting for others and writing books for a popular audience. He was not a member of the elite circles in which Darwin moved, although later he was befriended by Darwin, who petitioned the government to award Wallace an annual pension, which they did in 1880. Without the time, money, connections, and social capital, Wallace was unable to take the lead in scientific discussions about evolution, which became associated almost exclusively with Darwin's name rather than his own.

Wallace was born into a precarious middle-class family (his father Thomas had studied law but never practiced) and was only able to attend grammar school until he was 14. After that, he apprenticed to become a surveyor and worked in a number of surveying positions until his twenties. While working, he became interested in biology and voyages of scientific discovery. He read Malthus, Lyell, Chambers, and Darwin's account of his time on the *Beagle*. He also befriended the entomologist Henry Walter Bates (1825–92), with whom he took his first voyage of exploration and collection to South America in 1848.

Russel Wallace (1823–1913), announced in a letter to Darwin that he had developed a theory of evolution by natural selection and thus forced Darwin's hand. Darwin asked Lyell for advice, and Lyell obligingly delayed publishing Wallace's paper until Darwin had quickly written a paper of his own, so that a joint paper could be read before the Linnaean Society. Within a year, Darwin wrote a full statement of his theory in *On the Origin of Species by Means of Natural Selection* (1859).

There were a number of similarities in the paths Wallace and Darwin took. Both had traveled on collecting voyages, both had been inspired first by Lyell and then by Malthus, and both were interested in the distribution of species, that is, the question of why species occur geographically nearest to their closest relations. Moreover, both were involved in the imperial project, Darwin as a member of the industrial gentlemanly class, and Wallace as a paid collector, traveling to Amazonia

Bates and Wallace spent two years exploring the Amazon River, and a further four years on the Rio Negro. There they found an astonishing wealth of new species of plants and animals never seen by Europeans. Wallace had hoped to pay for this trip through selling his collection on their return, but unfortunately, after 26 days at sea the ship caught fire and burned to the waterline. Wallace was only able to save a few notes and sketches; everything else perished.

From 1854–62, Wallace traveled through the Malay Archipelago, collecting specimens to sell and studying the natural history of the region. He noted that there was a pronounced difference in species distribution on different sides of the archipelago. This made a strong case for the historical division of species, based on geographical distribution (a case later made by Darwin). This distribution line in the Malay Archipelago is now called the Wallace Line. He collected over 126,000 specimens, thousands of which had been unknown to science until that time.

While there he had his insight into the mechanism of evolution (natural selection), which he outlined in a letter to Darwin in 1858, thereby forcing Darwin to get his better-articulated theory into print the next year.

On his return to England, Wallace established himself as a successful writer. His most famous book, *The Malay Archipelago* (1869), has never been out of print. He began to associate with the luminaries of natural history and he met and conversed regularly with Darwin. But his finances were always precarious. He lived from book advance to book advance. Further, his interest in social issues (inspired by Robert Owen) and his foray into spiritualism isolated him from the gentlemen fellows of the Royal Society. Wallace never achieved the fame of Darwin and largely became a sort of footnote to Darwin's story. His lack of gentlemanly connections and resources contributed to his lower status in the scientific community.

and later to the Malay Archipelago. So perhaps it is not altogether surprising that both developed parallel theories on evolution by natural selection. Wallace always claimed that his was the less well-developed theory, and when he wrote his complete book on the subject, he called it *Darwinism* (1889). But the difference in their status, both scientifically and socially, also accounts for this difference in reputation. For good or ill, this theory has ever since been called Darwinian evolution.

Herbert Spencer and Social Darwinism

Darwin's theory struck a chord with educated Britons because, in many ways, it corresponded to social theories that had already been articulated. Evolution by natural selection fit especially well with Herbert Spencer's (1820–1903) ideas and become a major explanation for social growth and development. Widely popular, social Darwinism justified the belief in progress on one hand and the need to manipulate nature and society to achieve this progress on the other. This collection of theories taught people that evolution happened whether it was desired or not and, therefore, that civilized people were obligated, once conscious of it, to use this evolutionary force for good.

Herbert Spencer was a Quaker by birth, and after an early career as a railway engineer, he worked as a writer and sub-editor for *The Economist*. A small inheritance allowed him to devote his life to studying and writing, especially about the human condition. In such books as *Social Statics* (1851) and *Principles of Psychology* (1855), Spencer argued that the development of human societies could be explained as an evolution from simplest (native tribes) to most complex (European imperialist states). Like Darwin, Spencer was influenced by Malthus and saw life as a struggle in which only the strong tended to survive. He coined the phrase "survival of the fittest" to explain this phenomenon. To ensure the strongest and best society, he argued that it was desirable to keep state control to a minimum. Spencer did not support a completely individualistic morality, however, but maintained that strong communities were based on understanding the natural unity of interests of individuals. Survival of the fittest was thus key to the economic and social progress of mankind. These theories, when applied to race, were used as a justification for segregation (to keep the weaker races from mixing with the stronger) and subjugation (the strong had a moral and natural imperative to control the weak).

Spencerianism and forms of social Darwinism provided justifications for capitalism, laissez-faire economics, and arguments against a welfare state. It should come as no surprise that one of Spencer's biggest fans was the American industrialist Andrew Carnegie.

Other forms of social Darwinism stressed the struggle between races or nations, thus justifying the military and industrial competition taking place in Europe. These theorists believed that war was a way of winnowing out inferior nations. "Might is right" provided a justification for imperialism, since, according to its rationale, inferior races deserved to be exploited and controlled. Social Darwinism also had an individual expression in eugenics, which was presented by its proponents as the science of race. Eugenists claimed that the state had a duty to limit the multiplication of its least-fit citizens and to encourage its most fit to increase. An early proponent of this was Francis Galton (1822–1911), a cousin of

Charles Darwin, who wrote *Hereditary Genius* in 1869.

Darwin had not mentioned human beings in *Origin of Species*, although he was certainly aware that his theory could be so applied. At the 1860 meeting of the BAAS, Darwin's theories were attacked on this basis by Bishop Samuel Wilberforce, known as "Soapy Sam" in the press, and defended by "Darwin's Bulldog," Thomas Henry Huxley (1825–95), who replied that he would rather be descended from an ape than from an intelligent man who used his intellect to retard the growth of science! Darwin later entered the fray with the publication of *The Descent of Man* (1871), in which he set out to prove that evolution of intelligence and morality were also possible through natural selection and, therefore, that humans had evolved in the same way as had animals. This book lacked the rigor of his earlier work but produced a curious research avenue in which people examined their companion animals for signs of intelligence and humor. Wallace objected to this demotion of humans to animals and spent the last years of his life investigating spiritualism, a popular Victorian pursuit, in search of the divine spark that separated humankind from the beasts and that remained even after death.

PEDIGREE OF MAN

7.7 PEDIGREE OF MAN

A "family tree" based on an anthropocentric interpretation of Darwin's theory of evolution. This schema presents humans as the ultimate product of evolution.

Although Darwin's books and theories were extremely popular – *Origin* sold out within days of the first printing – his theory of evolution by natural selection did not win over the biological community before the twentieth century. He certainly had his contemporary supporters, especially Huxley in Britain and Asa

Gray (1810–88) in the United States. Naturalists were very favorably disposed to his theory, since it explained both why classification worked and the source of geographical distribution of species. Naturalist Henry Walter Bates, for example, argued that natural selection could be seen in the Midlands moth. Prior to the industrialization of the Midlands, the moths were predominantly light-colored to blend with tree bark and hide from predators. Due to pollution, trees had become darker, and dark moths gained camouflage advantage and multiplied, becoming the dominant form. Bates also put forward a Darwinian explanation of mimicry, the phenomenon of a harmless insect looking like a specimen of a poisonous species. As Bates pointed out, an insect that looked unpalatable was more likely to survive, and so such coloring could be a variation selected for survival.

Opposition to Darwin's Theory

Most naturalists, especially in Britain, were less concerned with evolutionary theory and more interested in collecting. A natural history collecting mania had taken hold, encouraging men, and increasingly women, to tramp the land in search of rare plants, animals, insects, and fossils. Middle- and upper-class collectors paid others, like Wallace and Bates, to travel to exotic locations and bring back specimens. Many of these amateur collectors were deeply influenced by natural theology, a doctrine articulated most clearly by William Paley (1743–1805). Paley argued that if a man were walking in the woods and found a watch, even if he had no idea what it was or how it worked, he would know that it had been made by some intelligence, that is, by the watchmaker. How much clearer it was, then, that the natural world, more intricately constructed than the watch, must have been fashioned by the Great Watchmaker. This argument, called the argument from design, was a potent one, which was later taken up by the authors of the *Bridgewater Treatises*. This series of books, commissioned by a bequest from Francis Henry Egerton, Earl of Bridgewater, attempted to link various aspects of science to a proof of God's existence and a divine plan, with William Whewell writing the treatise on astronomy and physics and William Buckland the one on geology. Darwin himself was careful not to contradict this idea of design, although many of his detractors saw atheism in the heart of his system, since it seemed to require no final plan to create humans.

While evolution as a basic premise steadily gained adherents in the scientific world, Darwin's theory did raise questions. How could God create a world that was so violent, so wasteful, and so unlike the grand design envisaged by the natural theologians? Could the human species really be descended from apes and not created in God's image? If Darwin was right, did that mean parts of the Bible were wrong?

For biologists and other scientists, other pressing issues were also unresolved. The fossil record did not seem to contain any gradual evolutionary forms, and there were many "missing links," or unfound intermediate forms, predicted by Darwin's theory. Present-day variations seemed too small to have created evolutionary change of the magnitude of the fossil record. More damning, the socially powerful and influential physicist William Thomson (1824–1907), Baron Kelvin of Largs or more simply Lord Kelvin, suggested that the age of the planet must be between 20 million and 400 million years, too short a time for Darwinian gradual evolution.

Thomson was a prodigy, accepted into the University of Glasgow at the age of ten, then moving to Cambridge to complete his education. He was made professor of natural philosophy at the University of Glasgow in 1846, holding that position until his retirement in 1895. He did groundbreaking work on electricity and magnetism and helped to lay the first Atlantic telegraph cable. For his work, he was knighted in 1866 and made a peer of the realm in 1892. He was also president of the Royal Society from 1890 to 1895. Lord Kelvin, basing his conclusions about the age of the Earth on its temperature and on the rate of cooling of the planet from a molten state, argued that its age was about 50 million years. This calculation came from his work on thermodynamics, particularly his *On the Dynamical Theory of Heat* (1851), in which he introduced the absolute (or Kelvin) scale of temperature, setting zero at the theoretical point at which molecular motion stopped. On this absolute scale, water melts at 273.16°K. Assuming that the planet was moving along a heat spectrum from molten to the temperature of space (i.e., it is a rock heated only by the energy of the Sun), the time that life could have existed on Earth could be calculated. The problem for Darwinian evolution was that 50 million years, while a very long time, was a far cry from the 20,000 million years that some scientists had suggested were necessary for evolution to produce life as it now existed.

Finally, there were questions about the biological mechanism by which variations or inherited characteristics were passed on. Darwin's speculation on the mechanics of reproduction were not well received. Combined with other objections, many scientists were reluctant to accept his explanation. By 1900, while evolution was part of every biologist's credo, the mechanism was more in doubt than ever.

CONNECTIONS

The "Warfare" between Science and Religion

In 1875, John William Draper (1811–82) wrote *History of the Conflict between Religion and Science*. In this book, he argued that religion (by which he meant Christianity and often Catholicism) had held back scientific progress for many centuries and that it was only now, in the enlightened nineteenth century, that scientific thinking was able to prevail over religious superstition. Draper argued that the history of science had been a history of conflict between science and religion, with science seeking to expand the power of human intellectual prowess and religion seeking to stifle such ability. He cited the medieval suppression of Greek knowledge of the sphericity of the Earth and the Roman Catholic Church's persecution of Galileo as examples of this warfare. This thesis (the "conflict" thesis) became a powerful and often-cited explanation and continues to be used to the present day, especially concerning the relationship between evolution and creation.

The reality, however, is much more complicated. First, the terms "science" and "religion" themselves have changed over time. For many millennia, as we have seen, science and religion were closely intertwined and many natural philosophers would not even understand the distinction Draper was trying to make. The relationship between understanding God's world and the path to salvation was often a positive one, rather than in any way confrontational. Second, the specific examples Draper used were not accurate representations of this relationship. With regards to the flat-earth

The Professionalization of Science and Science Education

It is not surprising that the era that shaped Darwinian theory was also the era that produced Sherlock Holmes, Sir Arthur Conan Doyle's cool, scientific, and supremely logical private detective. Scientists, even if they continued to argue the fine details, seemed to be revealing all of nature, and it seemed possible to create a hypothesis, there was no suppression of spherical interpretations. The consensus among Christian scholars and most vernacular writers interested in the physical shape of the Earth, from the fall of Rome to the years before Columbus, was that the Earth was round. There was no viable dissenting movement claiming the world was flat and no suppression of the idea of a globe earth by religious authorities.

Likewise, the conflict between Galileo and the church was less about religious doctrine and more about political conflicts within the Catholic church. Galileo was a staunch Christian, and worked to reconcile science and religion. He hoped that good science would help the Catholic church.

Before the nineteenth century, the vast majority of scientists were believers in a variety of religious faiths. They saw their religion as an aid to scientific investigation, acting as a sanction for pursuing science, providing them with starting hypotheses, and giving them design arguments, for example. This began to change in the mid-nineteenth century. Auguste Comte's (1798–1857) positivist philosophy argued that science displaced religion in an onward march of progress. William Whewell, the British philosopher of science, declared in his *History of the Inductive Sciences* (1837) that science must be released from former superstitious trappings in order to flourish. Draper's book, and a subsequent book by Andrew Dickson White (1832–1918), *A History of the Warfare of Science and Theology in Christendom* (1896), continued this trend. By the 1870s, the move to professionalize science made many argue that science was a different enterprise than religion (although not necessarily antagonistic to religion), especially in Britain.

Religion and science, however, have remained in a complex and often mutually beneficial relationship throughout the twentieth and into the twenty-first century. There continue to be many scientists who are believers, some of whom compartmentalize these endeavors and some of whom are influenced by their beliefs. Equally, there are religious individuals who see science as an important element of the created world. The relationship between these two knowledge systems is complex and in constant reinvention.

complete picture of how the physical world operated. With such knowledge, no secret could be kept from the observant mind. Darwin and Sherlock Holmes were also alike in being amateurs rather than professionally trained and employed. While Holmes gave metaphorical birth to generations of private detectives, Darwin, as gifted and insightful as he was, represented the waning of a scientific style. He was the last of the great amateur gentleman scientists. For science as an occupation, the nineteenth century was a turning point. There was an increasing sense of science as a professional activity, and with this came a growing separation of the study of nature as a branch of philosophy. This was spurred by the appearance of more and increasingly specialized scientific organizations and the development of new institutions in education and research dedicated to science. While it may have been used by German academics earlier, the term "scientist" was a product of the nineteenth century, introduced by William Whewell (1794–1866) to a large audience at a meeting of the BAAS in 1833.

The reign of Napoleon offers a good starting point for the expansion of science as a profession. Although Napoleon's actual support for science was somewhat variable (for example, he abandoned a number of scientists in Egypt when he escaped the British there), he did foster a sense of the importance of technical knowledge. The Académie des Sciences was disbanded in 1793 during the Revolution, but it re-emerged in 1795 as part of the Institut National. In addition to the Académie Napoleon approved the foundation of the Société d'Arcueil in 1805, which had as members such leading thinkers as Claude Berthollet, Pierre-Simon Laplace, and Alexander von Humboldt.

More important than the Société d'Arcueil was the transformation of the École Polytechnique into the leading scientific and engineering school of the era. Originally founded as the École centrale des travaux publics in 1794 by the National Convention, it changed its name in 1795 and absorbed the state artillery school in 1802. First under the Ministry of the Interior, the connection with the military was completed in 1804 when Napoleon transformed it into an elite military school. The artillery had always been the most intellectual branch of the military, requiring a broad understanding of mathematics, physics, chemistry, and elements of what we would call materials science. At the same time, the artillery was not steeped in ancient traditions and offered a path to military command that was not determined solely by social rank. It had attracted many bright young men of lower birth, including Napoleon, who was commissioned as an artillery officer in 1785.

The École Polytechnique was founded by three men: Lazare Carnot (1753–1823), who organized the Republican armies and wrote on the science of fortification; Gaspard Monge (1746–1818), a mathematician and physicist, whose work on descriptive geometry laid the foundation for architectural and engineering drawing; and Adrien Marie Legendre (1752–1833), a mathematician who worked on number theory. Mathematics, chemistry, and physics were core subjects, and the material was often so advanced that there were complaints that the École Polytechnique was too theoretical to be useful as a military school. Yet its reputation was so great that it continued to function throughout the century and continues as one of France's leading educational institutions today. It has produced a stream of professional engineers and scientists for more than 200 years.

The grand paradox of Napoleon's reign was that he ended the Republic and failed to conquer Europe but set the continent on a course that brought about the very reforms that other countries had gone to war against him to defeat. With the exception of the British, who had suffered through similar reforms a century earlier, the wars forced many European countries to undergo major economic and social change. Agrarian economies and restricted middle classes could not provide the material resources to counter the Napoleonic threat. To match the power of France, the continental powers needed the support and productivity of their citizens, and in turn those citizens demanded more autonomy and a greater say in government. Pre-Napoleonic armies were based on late Renaissance structures and commanded by the often-untrained nobility, but the death toll among officers and the rapid expansion of the size of Napoleonic-era armies forced a change in military command. Napoleon had taken French peasants and middle-class burghers and turned them into a powerful army. Many new officers, who were not drawn from the nobility, led forces into battle for or against him. They were not willing to go home and return to the old system.

The freedom and broad scope of the amateur scientist had benefited British science in the years from Newton to the early nineteenth century, but with the increasing complexity of scientific knowledge, that approach was failing. The centers of scientific strength shifted to France and, even more, to German schools and research organizations. In Britain the rhetoric of scientific utility as espoused by the Royal Society rang increasingly hollow. The Royal Society was doing less and less to support science and scientists and had, in fact, become little more than a kind of exclusive social club. Most members conducted little or no scientific work, yet they sat in judgment on scientific matters and offered advice to the government. Scientists such as Charles Babbage (1792–1871) were particularly concerned about the lack of government support for science (although he received a number of grants for his work on calculating machines) and the state of the Royal Society. Babbage wrote *Reflections on the Decline of Science in England* (1830) and published a second edition with additional material and a foreword by Michael Faraday in 1831. It was a vitriolic attack on the Royal Society. While many of its points were valid, it did little to persuade the Royal Society to change its orientation.

With the Royal Society seeing no reason to change, Babbage and a number of his friends opted to create a new society dedicated to science and the promotion of science. In 1831 they founded the BAAS, modeled on the Deutsche

Naturforscher-Versammlung, which had been created in 1822 by Lorenz Oken (1777–1851). Where the Royal Society and the Académie des Sciences were elite organizations that treated science as a superior intellectual pursuit and in which membership was tightly restricted, the new societies were organized around actual participation in science. The BAAS held meetings not only in London but all over the country and in colonial territories such as Canada. These meetings encouraged scores of naturalists by providing a venue for the presentation of local knowledge that was ignored or deemed too insignificant for the elite organization. The BAAS also developed closer ties with industry than did the Royal Society, since its membership was open to anyone interested in science including small-business owners, schoolteachers, and craftspeople, unlike the Royal Society, which continued to select members by sponsorship and election.

Babbage and his supporters were concerned about Britain losing its leading position in science, but for the most part the consequences of falling behind went unnoticed, masked as they were by the growing power of the British Empire. In 1851 the British celebrated that power with a massive display of technology and empire at the Great International Exhibition in London. The centerpiece of the Exhibition was Joseph Paxton's Crystal Palace, a display hall built of a cast iron frame and glass panels. Inside were the industrial marvels of the day, a cornucopia of machines and products brought together to demonstrate and promote technical innovation. The Exhibition was a massive undertaking, with 13,000 exhibits from around the world including industrial and commercial displays and fine arts. It also included exotic displays from the empire, with plants such as a mammoth water lily from British Guiana that had leaves large enough to support a child's weight. The Exhibition attracted over 6 million visitors. The building itself was an innovation, a forerunner of modern modular steel-frame construction. Paxton used 4,000 tons of iron for the skeleton and 83,610 m² of glass exterior skin to enclose 71,800 m² of space in Hyde Park. (See figure 7.8.)

Associated with the Exhibition were efforts to educate people about and to popularize fine arts, business, and science. A number of people noted that, while Britain led the world with

7.8 CRYSTAL PALACE

The Crystal Palace, the central exhibition hall of the Great Exhibition of 1851.

her industrial power, she lagged behind her rivals in training and institutional support for technical and scientific education. In the same year as the Exhibition, the Government School of Mines and of Science Applied to the Arts opened its doors, a partial response to the continuing concern about science training. The school was renamed the Royal School of Mines in 1863, and it added the "New Science School" in 1872; it was built in South Kensington with profits from the Exhibition. The New Science School contained the Departments of Natural History and Physical Science. While these efforts helped Britain, they did not match the support and scope of technical and science training in France or Germany.

Louis Pasteur

In France, the École Normale Supérieure produced generations of important scientists, including people such as mathematician Évarist Galois (1811–32), sociologist Émile Durkheim (1858–1917), and biologist Louis Pasteur (1822–95). Pasteur, who was broadly trained in chemistry, physics, and biology, began his scientific career investigating the asymmetrical crystalline structures of acids. As a vitalist, he was convinced that living organisms were intrinsically different from nonliving matter. As significant as his crystallography was, he is most famous for his bacteriological work. He discovered that the fermentation process – in wine, milk, and vinegar – was due to the activity of microscopic animals rather than a chemical reaction. Carrying on the observations of seventeenth- and eighteenth-century microscopists, and adding carefully controlled experimental work, Pasteur discovered that these micro-organisms were anaerobic (lived without oxygen) and that they could be killed by heat. The process of pasteurization, used in winemaking (and later in milk production) to kill harmful microbes, boosted the French wine industry, as well as earning Pasteur a good return on his patents.

During his career, Pasteur extended the germ theory of disease, first put forward by Robert Koch (1843–1910), to diseases in silkworms; developed vaccinations for anthrax and rabies; and helped establish Pasteur Institutes around the world to carry on his research. He was a skilled self-promoter and won an important scientific debate with Felix Pouchet (1800–72) concerning the possibility of spontaneous generation. Pouchet believed that life could be created from inanimate matter under the right conditions (such as warm, moist earth or dung). When an experiment proved inconclusive, Pasteur simply said that spontaneous generation was not possible, and his status as the most famous French scientist of the day ended the argument. Later work confirmed Pasteur's position, but spontaneous generation largely disappeared as idea in biology.

Science in Japan: The Fusion of Ideas in the Global Context

By the beginning of the nineteenth century, colonialism and global contacts had made clear to many people, often in the bloodiest terms, the power of European technology. Some non-European nations began to realize that there was a scientific foundation to this imperial power and began to seek it out for themselves. In some cases, scientific exchange was well established; for example, the Jesuits had been teaching in China since the sixteenth century. In the early days, the Jesuits learned more from the Chinese than the Chinese learned from the Europeans (a situation repeated in the McCartney mission of 1793), but by the nineteenth century the scientific ideas of Europe took on new importance for non-European powers with the rise of European imperialism.

In the case of Japan, the transformation of a small group of scholars and collectors creating an Indigenous natural philosophy into a scientific community was spurred first by interest in Chinese scholarship and later by European ideas, sometimes smuggled into Japan. During the Tokugawa period (1600–1868), astronomers and teachers were employed by the shogun and received a modest salary. Physicians, who were often trained in natural philosophy as well as medicine, had a higher social standing than astronomers or philosophers. Prior to the *sakoku* (closed country) policy of 1633–9, Dutch and Portuguese traders brought European medical texts to Japan. Information about surgery and *materia medica* (pharmacology) were particularly sought after because of their utility. After 1639, the only foreign traders allowed in Japan were the Dutch, Korean, and Chinese, all strictly controlled. In 1650, shogun officials ordered a European anatomical textbook from the Dutch and directed a number of physicians to study Western medicine. Over time, the Japanese examined and translated a number of European medical books and growing interest in experimental physiology and anatomy following the European style led to a decline in Chinese-based medical practice.

European astronomy (primarily Ptolemy and its Aristotelian foundation) had been introduced by Jesuits prior to the *sakoku* policy, but the introduction of the Chinese Shou-shih calendar around 1670 offered a much more practical approach to astronomy. This was not just rote adoption, since the Chinese calendar was modified for Japan using a European world map to adjust for the different longitude and latitude. The fusion of ideas can be clearly seen in the life and work of the astronomer Goryu Asada (1734–99). Trained as a physician, he taught himself astronomy and left his clan to do astronomy in Osaka. Shigetomi Hazama (1756–1816), his patron and student, owned a copy of a Chinese astronomical text that had been edited by a Jesuit who included Kepler's three laws. According to the historian of science Takehiko Hashimoto, the Keplerian agreement between observation and theory impressed Asada so much that he began to study Western astronomy. In turn, the skills learned from European astronomy led to a mathematical survey of Japan and the creation of the first indigenous map of Japan in 1821.

One of the crucial changes in the place of science in Japan came about because of the entry of the samurai class into "technical schools," which taught subjects such as astronomy, physics, and mathematics. This increased both the interest in and status of the physical sciences. Interest in Western science was further increased by the arrival of Commodore Perry in 1853. Perry's arrival forced the Japanese to open diplomatic relations with the United States and the appearance of his ships made the Japanese realize that the technology of the Americans was far ahead of anything in Japan. By 1870, interest in Western practices prompted the Higo administration to close the Jishukan academy, a Confucian school, and open the Yogakko academy run by Captain L.L. Janes, a graduate of West Point. In addition to English instruction, he taught mathematics, chemistry, physics, and geology.

In 1868 the Tokugawa shogunate ended and the Meiji ("enlightened rule") period began. This was characterized by restoration of the power of the emperor, but also by the Charter Oath:

1. Deliberative assemblies shall be widely established and all matters decided by open discussion.
2. All classes, high and low, shall be united in vigorously carrying out the administration of affairs of state.
3. The common people, no less than the civil and military officials, shall all be allowed to pursue their own calling so that there may be no discontent.
4. Evil customs of the past shall be broken off and everything based upon the just laws of Nature.
5. Knowledge shall be sought throughout the world so as to strengthen the foundation of imperial rule.

While clause five clearly indicated Japan's desire to select the best ideas and practices, including those of science, from anywhere in the world, it is also important to note that clause four contains the same conception of natural law as was promoted by many European thinkers in the late eighteenth century.

Japan provides an excellent example of the importance of science to modernization for non-Western countries. Although this science originated in Europe, it was not a Western idea that was somehow imposed on others by colonialism or other forms of coercion. Japanese scholars and leaders had a variety of sources for natural philosophy and in the beginning used a combination of Chinese and domestic ideas as the foundation of Japanese natural philosophy. When competing ideas filtered in from Europe, they accepted or rejected them based on their utility, but they also blended and adapted those ideas. Today, Japan is one of the most science-oriented countries in the world, but it retains its own traditions and culture.

Chemistry and the State: Classification, Structure, and Utility

While French science prospered through the work of scientists such as Pasteur, Germany was creating a scientific program that proved incredibly fruitful. By 1850, the German states had promoted a strong scientific culture, transforming teaching and research at the universities and creating a network of partnerships among scientists, business, and government. At first, Britain, blinded by wealth and empire, did not recognize the extent of Germany's progress and relied on the eminence and expertise of such established scientists as Lord Kelvin. Although some British chemists and physicists were on a par with French and German researchers in terms of "pure" research, the continental scientists were supported by the money and effort poured into science and engineering as part of a larger integration of science with state and industrial needs. Chemistry had emerged as a separate category of research in the eighteenth century, but in the nineteenth century it was in a state of flux. Much industrial chemistry was based on a craft or artisan system of production, but by the middle of the nineteenth century sharply increased demands for materials led to a greater need for mass production methods. Some products such as gunpowder, dyestuffs, acids, and naval materials such as pitch were so important and in such high demand that they became issues of national security. To understand how obtaining and maintaining supplies of these materials would develop into a serious national issue, we have to head back to the laboratory and pick up some of the threads of chemistry left from the previous century.

At the turn of the century new processes were being pioneered, and more substances were being rapidly created and discovered. While this demonstrated a robust interest in chemistry and helped supply industry, it also meant there was more to be confused about. For example, Lavoisier had listed 33 elements in his work *Elements of Chemistry*, but 32 new elements were added to the list by 1860, as well as 1,000 compounds that had not existed in the previous century. Chemists were developing better laboratory techniques, made possible by new tools such as the Bunsen burner, introduced by Robert Bunsen (1811–99) and his student Henry Roscoe (1833–1915). The burner's colorless flame heated substances to a point hot enough to emit light. As a result, chemists found that each substance had a unique pattern of spectral colors, a finding that proved extremely useful for analytical work, including Bunsen's discovery of two elements: rubidium and cesium.

Although better tools and techniques provided interesting and useful materials, fundamental questions about chemical activity required an overarching theory to pull the information into a comprehensive system. Order was needed, and some felt that it could be obtained by looking for the hidden relationship of the elements. John Dalton (1766–1844) had propounded an atomic theory that distinguished elements by their atomic weights (relative masses based on the assignment of a comparison of elements with hydrogen as the lightest). This idea was published in his *New System of Chemical Philosophy* (1808) and widely influenced how matter was understood. Yet a simple linear list of elements by weight did not provide much structure. Around 1829 Johann Döbereiner (1780–1849) put elements in groups of three, called "triads," and noted similar properties in the grouped elements. In 1862 A.E. Béguyer de Chancourtois (1820–86) published what he called the "telleric helix" that placed the known elements on 45° lines arranged on a cylinder. It was a step toward organizing the elements by weight and characteristics, but it gained little notice.

In 1826 the great chemist Jöns Jacob Berzelius (1779–1848) calculated atomic weights for 48 elements and molecular weights for hundreds of associated oxides. His work was not universally accepted, however, as other researchers proposed alternative atomic weights and systems of measurement. Different scientists used different standards for atomic weight for their measurements, some using hydrogen as one on the scale, while Berzelius set oxygen as equal to 100 and measured weights against it. After several years of work, in 1860 Stanislao Cannizzaro (1826–1910) announced a new way to deal with the problem of measuring atomic weight when he demonstrated that the density of gases and vapors could be compared with the density of hydrogen in order to determine the molecular weights of elements and compounds accurately.

With a reliable method for determining atomic weight in hand, it became possible to address the question of classification of the elements. Julius Lothar Meyer created a table of elements in 1864 that placed elements with similar properties in columns in sequence with their atomic weights. These analogous elements also exhibited similar valence. Valence was initially a measure of the combining power of an element. Elements with a high valence, such as oxygen, vigorously combined with other elements to form compounds, while other elements, such as gold, only rarely formed compounds. In modern terms valence is based on the number of hydrogen atoms with which an element may combine; for example, one atom of oxygen can combine with two atoms of hydrogen or one other atom of oxygen. This system also correlated with the chemical activity of elements, not just quantity. For example, the elements with valence of 1 (such as alkali metals) were very reactive, while elements with valence of 3 (nitrogen or arsenic) were much less reactive. The four most important elements for organic chemistry form a nice list by valence:

H (hydrogen)   1
O (oxygen)   2
N (nitrogen)   3
C (carbon)   4

In Britain, analytical chemist Alexander Reina Newlands (1837–98) independently produced a table of elements that grouped them by families and atomic weight. When he presented his work to the Chemical Society in 1866, it was criticized both for problems with the analogous groups and because it left no spaces for the apparent gaps or odd jumps in weight that seemed to exist within various groups. His paper "The Law of Octaves and the Causes of Numerical Relations between Atomic Weights" was rejected by the *Journal of the Chemical Society*. This rejection came back to haunt the Chemical Society when the periodic table, which looked very much like Newlands's work, was established not by a British chemist but by a Russian. The Chemical Society belatedly awarded Newlands its highest prize, the Davy Medal, in 1887.

Mendeleev and the Periodic Table

Julius Lothar Meyer (1830–95) and Dmitri Ivanovitch Mendeleev (1834–1907) gave order to the elements. Each independently recognized that the elements could be grouped by both atomic weight and by characteristic. In 1868 Meyer created nine columns and placed elements with similar characteristics in ascending order of atomic weight. That same year Mendeleev began to write a chemistry textbook

Reihen	Gruppe I. — R²O	Gruppe II. — RO	Gruppe III. — R²O³	Gruppe IV. RH⁴ RO²	Gruppe V. RH³ R²O⁵	Gruppe VI. RH² RO³	Gruppe VII. RH R²O⁷	Gruppe VIII. — RO⁴
1	H=1							
2	Li=7	Be=9,4	B=11	C=12	N=14	O=16	F=19	
3	Na=23	Mg=24	Al=27,3	Si=28	P=31	S=32	Cl=35,5	
4	K=39	Ca=40	—=44	Ti=48	V=51	Cr=52	Mn=55	Fe=56, Co=59, Ni=59, Cu=63.
5	(Cu=63)	Zn=65	—=68	—=72	As=75	Se=78	Br=80	
6	Rb=85	Sr=87	?Yt=88	Zr=90	Nb=94	Mo=96	—=100	Ru=104, Rh=104, Pd=106, Ag=108.
7	(Ag=108)	Cd=112	In=113	Sn=118	Sb=122	Te=125	J=127	
8	Cs=133	Ba=137	?Di=138	?Ce=140	—	—	—	—
9	(—)	—	—	—	—	—	—	
10	—	—	?Er=178	?La=180	Ta=182	W=184	—	Os=195, Ir=197, Pt=198, Au=199.
11	(Au=199)	Hg=200	Tl=204	Pb=207	Bi=208	—	—	
12	—	—	—	Th=231	—	U=240	—	— — —

7.9 MENDELEEV'S PERIODIC TABLE FROM *ANNALEN DER CHEMIE* (1871)

that was to be a great synthesis of chemical knowledge. Like Lavoisier did in the *Elements of Chemistry*, he attempted to classify all known elements. He made a card for each element, listing its properties, and then arranged and rearranged the cards, looking for some pattern. A pattern emerged when the cards were sorted by atomic weight. He produced a table of elements in 1869 and an improved version in 1871. (See figure 7.9.) While there continued to be problems, especially with some of the more recently discovered elements whose characteristics were not completely certain, his arrangement made clear both the progression of atomic weight and the collection of series by characteristics and valence.

One of Mendeleev's great insights was to leave gaps representing undiscovered elements, whose particular characteristics he then predicted. Mendeleev beat Meyer to press in 1869, but Meyer included some of Mendeleev's work when he produced a new table in 1870 that was almost identical to the modern periodic table of elements.

Mendeleev's system pulled together a vast array of chemical ideas and information. Just as Lavoisier's nomenclature contained an underlying philosophy within it, so too did the periodic table present a particular philosophy of matter. It treated matter as a collection of distinct and indivisible particles. The particles were capable of combining to form more complex compounds, but they could do so only in fixed proportions. Atoms of an element were identical and behaved in the same manner. The table also codified certain conceptions of the physical world such as the quantity of atoms in a given mass of material, valences, and the definition of elements. While the modern periodic table of elements tends to be

7.10 MENDELEEV'S PREDICTIONS AND
BOISBAUDRAN'S ANALYTICAL RESULTS

PROPERTY	EKA ALUMINUM	GALLIUM
Atomic weight	≈68	69.9
Density	5.9	5.93
Melting point	Low	30.1°C
Formula of oxide	Ea_2O_3	Ga_2O_3

treated as a kind of index, it has behind it a long history of debate over all these questions about the structure and condition of matter. When a scientist uses the periodic table, he or she accepts the philosophic ideas encoded in it. After long use and demonstrated reliability, the acceptance of the periodic table has become unconscious, so deeply imbedded in the practice of science that it is considered axiomatic and thus very hard to challenge.

The utility of Mendeleev's system was further supported by the discovery of the elements represented by the blank spaces in his table. Among others, Mendeleev predicted the properties of what he called *eka aluminum* (*eka* meaning first in Sanskrit). (See figure 7.10.) In 1875 Paul Émile Lecoq de Boisbaudran (1838–1912) discovered gallium, which had the basic characteristics Mendeleev predicted for eka aluminum. When his "ekaboron" was discovered in 1879 by L.F. Nilson (1840–99) and called scandium, Mendeleev's fame was confirmed.

For his achievements, Mendeleev won the Davy Medal with Meyer in 1882, was elected a Foreign Member of the Royal Society in 1890, and was awarded the Copley Medal in 1905.

The Structure of Organic Compounds

While the periodic table provided a powerful tool for understanding the material world, it also made clear a troubling problem about the nature of matter. With over 60 verified elements discovered by 1875, questions arose: Why were there so many different elements? Could small changes in mass really account for their vastly different characteristics? Once the pattern of valence was established, it raised the further question of how atoms stuck together. Affinity theory increasingly seemed inadequate, for it offered no physical explanation for why a carbon atom would have twice the affinity or combining power of oxygen.

Although elements were pure and relatively limited in number, the number of molecules that could be built up out of the building blocks was immense, especially in organic chemistry. Modern organic chemistry looks at any compound that has carbon as its central component, but many chemists in the eighteenth century thought that organic compounds could be produced only by living organisms. According to this theory, called "vitalism," there was a special force of life in plants and animals. This idea was based on the simple observation that plants and animals were alive and, although composed of chemical elements, clearly were different from inorganic matter. While many believed in vitalism, researchers did not articulate it clearly. Some saw vitalism as a special property of life that could be studied, while others saw it as the divine spark and, therefore, not subjectable to scientific investigation. Friedrich Wöhler (1800–82) struck a blow against the theory of vitalism when he produced urea from inorganic compounds in 1832, although it would take more than 100 years for vitalism to disappear entirely. By treating lead cyanate with ammonium hydroxide and removing lead oxide, Wöhler was left with urea – in modern nomenclature $CO(NH_2)_2$. This was a major step toward the manipulation and synthesis of organic compounds, one that had a fundamental role to play in the European race to create high-technology industry.

By the 1840s there had been a flurry of work on the synthesis of organic compounds and, more importantly, experiments on how the component parts combined and could be substituted. Much of this work was based on the theories of Berzelius, who introduced ideas about electrochemical combination and radicals. Radicals were groups of atoms that were a consistent base. They could have other atoms attached to them (usually from a small list of possible substitutes), which varied the properties of the compound but created a kind of family of related compounds.

Jean-Baptiste-André Dumas (1800–84) introduced his theory of types as a way of understanding the composition and activity of organic compounds by grouping them according to properties and reactions. In turn, his student, Auguste Laurent (1807–53) proposed the "nucleus theory": since the structure of the compound determined potential chemical reactions, the place and relationship of the atoms in a substance was vital to understanding chemical behavior. Each of these ideas, strongly argued for and against by their supporters and detractors, contributed to an understanding of the complex structure and behavior of organic compounds. What held back a generally accepted system of organic chemistry was partly a lack of coherent information, since, for example, different chemists used different atomic weights for the same atoms in a compound. This produced apparently different results from the same analysis and, hence, did little to verify the theories being used.

The problem was bonding, or the way various atoms joined together. Since differing quantities of atoms could be made to create molecules under different circumstances, it was unclear how the rules governing molecule creation actually worked. Friedrich August Kekulé (1829–96) helped to clarify this problem. Rather than seeing radicals as the functional component, each atom should be treated

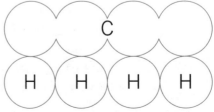

7.11 KEKULÉ'S SAUSAGE FORMULA FOR CH_4
One carbon atom is linked to four hydrogen atoms.

7.12 KEKULÉ'S ACETIC ACID TABLE

$C_4H_4O_4$	Empirical formula
$C_4H_3O_3 + HO$	Dualistic
$C_3H_3O_4H$	Hydrogen acid theory
$C_4H_4 + O_4$	Nucleus theory
$C_4H_3O_2 + HO_2$	Longechamp
$C_4H + H_3O_4$	Graham
$C_4H_3O_2O + HO$	Radical theory
$C_4H_3O_2 \atop H]O_2$	Gerhardt's type
$C_4H_3 \atop H]O_4$	Schischkoff's type
$C_2O_3 + C_2H_3 + HO$	Berzelius's copula
$HO(C_2H_3)C_2,O_3$	Kolbe
$HO(C_2H_3)C_2,O,O_2$	Kolbe
$C_2(C_2H_3)O_2 \atop H]O_2$	Wurtz
$C_2H_3(C_2O_2) \atop H]O_2$	Mendius
$C_2H_2HO \atop HO]C_2O_2$	Geuther
$C_2H_3 \atop O \atop O]O + HO$	Rochleder

as an equivalent part that could be combined following fixed rules. At the heart was carbon, which could combine with four other atoms (carbon was "tetratomic" or "tetrabasic" as Kekulé called it) but need not be limited by radical or family groups. In 1857 Kekulé represented the relationship with his "sausage formulae." (See figure 7.11.)

The emphasis on structural ideas was critical for organic chemistry, as Kekulé made clear in 1864 when he published a list of 20 different formulae for acetic acid from various chemists to demonstrate the extent of confusion. (See figure 7.12.) While there was general agreement on the empirical formula, $C_4H_4O_4$ (representing four carbon, four hydrogen, and four oxygen atoms), there was complete disarray about how to represent the components and what the various functional parts might be.

Following Kekulé's ideas, a graphic method of presenting the complex structures of organic molecules based on the number of bonds that each atom could form was worked out.[3] In 1866 the system of representing links by lines appeared and continues to be the standard method to the present day. Although models of molecules had been built prior to the rise of structural theories, the clarity of the new system encouraged model building, which had important consequences for a number of research programs, most spectacularly with the race for the structure of DNA. Although the psychological shift was subtle, the change from the old alchemical concept of transforming matter to the new chemical concept of constructing molecules made chemistry more amenable to research on commercially useful compounds.

The triumph of structuralist ideas was based not only on rationalizing the nomenclature of organic compounds, however. Structural illustrations made clear that compounds with the same empirical formula (that is, with the same number and kinds of atoms) could have very different structures, and further, that certain classes of molecules differed only by small shifts of location of atoms in a larger structure.

....

3. The fundamental organic components are carbon, oxygen, hydrogen, and nitrogen, with a host of others such as phosphorous, sulfur, and iron playing important roles in living cells. The mnemonic HONC lists the atoms by number of bonds each has H-1, O-2, N-3 and C-4.

One of the greatest breakthroughs that came out of the structural approach was Kekulé's discovery of the benzene ring. Benzene and related substances, collectively known as aromatics because of their distinctive and sometimes pungent odors, were first noted by Michael Faraday in 1825. They were found in coal tar, a by-product of coke production, and had interesting properties such as the ability to combine with a wide range of other organic compounds. Benzene also presented a mystery, since its empirical formula C_6H_6 seemed to leave two leftover links that were not filled by hydrogen atoms. If the valence rules were correct, there were either two missing hydrogen atoms or four carbon atoms if the molecule had valences of five. Either possibility would require a rewriting of the definitions of atomic bonding and elements.

7.13 BENZENE RING

According to Kekulé's own account, he found a solution to the problem while relaxing by the fire. He imagined atoms in long rows and moving like snakes. One of the snakes seized its own tail in its mouth. Kekulé used this image to work out a hexagonal model for benzene with double bonds for pairs of carbon atoms and single bonds between the pairs. (See figure 7.13.) Although this story may have been a *post hoc* explanation (some historians have questioned Kekulé's account), it has become one of the icons of the history of chemistry.

Chemistry, Industry, and the State: The Creation of Synthetic Materials

By the beginning of the second half of the nineteenth century, German chemical industries were turning to research like Kekulé's to improve production and create new products, since the German states were far less well endowed with domestic natural resources than France or Russia and had far less access to colonial resources than Britain. Although unification in 1871 improved some aspects of domestic access to resources, Germany was very late entering the colonial game. This had major consequences later, as Germany embarked on a series of wars to rectify the situation. In the meantime lack of natural resources was a major impetus to German interest in science education and research.

In 1837 the German chemist Justus von Liebig (1803–73), one of the most influential scientists of the era, told the British Association that Britain was no longer a leader in science. When he toured that country in the early 1840s lecturing on various topics, he drove the point home by describing the new work being done in chemistry in Germany and the increasing importance of chemical education, especially laboratory-based training. Thus, the rise of German chemistry and the shift in the nature and locale of research did not go completely unnoticed in Britain. Liebig's own teaching laboratory at Giessen was a model of chemical education that combined intellectual training with practical laboratory work. It involved students in real research projects, giving them more than a technical overview of existing work.

Because there was no comparable chemical instruction at any institution in Britain, a group of prominent people, headed by Albert, the Prince Consort, and including Sir James Clark (the Queen's physician), Michael Faraday, and Prime Minister Sir Robert Peel, contributed to a fund to establish the Royal College of Chemistry. It opened its doors in 1845, under the leadership of August von Hofmann (1818–92), who had been recommended by Liebig and personally persuaded by the Prince Consort to come to Britain from Germany. Although the Royal College of Chemistry was a small step toward reviving science leadership in Britain, there was no British equivalent of the French École Polytechnique or École Normale Supérieure. Even British universities lagged far behind continental counterparts such as the University of Göttingen in the development of science programs.

The Royal College of Chemistry had as its principal aim the training of pure researchers, copying the high end of intellectual activity at continental schools, but did little to encourage any integration of research with application. Applied chemistry was, if not exactly discouraged, looked down upon as a rather second-class use of talent. It was certainly the case that many students and working chemists felt that any excursion into applied – or worse, commercial – chemistry could end a research career. Against this background, William Perkin (1838–1907) entered the College at the age of 15. Perkin's work opened the door to a whole new era in chemistry, but it was not the British who saw its importance.

High on the list of natural materials that many were attempting to synthesize was quinine, which, as we saw earlier in this chapter (see p. 223), was the only known treatment for malaria and was produced only from the bark of the cinchona tree, native to South America and difficult to cultivate elsewhere. By 1852 the East India Company alone was spending about £100,000 annually on quinine.

Hofmann believed that naphthalidine could be converted to quinine because it shared some of the same basic components. Naphthalidine was readily available because one of its components, naphtha, was a by-product of the production of coal gas, which was widely used for lighting and heat and was made by heating coal in the absence of air. The captured gas was about 50 per cent hydrogen and

35 per cent methane, with a mixture of other gases making up the remainder. What was left of the coal after the process of gas extraction was coke, which was used as a solid fuel, and a viscous brown-black tar. Coal tar was a rich source of over 200 organic compounds, including benzene, naphthalene, and toluene, which were useful in both research and commercial applications.

Perkin began his work on quinine in 1856. He created a small laboratory in his home and attempted to follow Hofmann's idea about the conversion of naphthalidine to quinine, but instead of quinine, the experiment produced sludge. However, he noticed that a rag used to clean up some spills of the failed experiment was stained an intense purple. Rather than seeing his experiment as a failure, he proceeded to purify and test this unknown substance, which turned out to be artificial aniline, the basic coloring component of the indigo plant. The intense color led Perkin to believe that he had created a product with commercial potential. With funding from his family, he established the first artificial dye plant, producing a line of mauve and purples. His endeavor was not an immediate success, however. The dyeing industries, which used materials from natural sources – plants such as indigo, madder, and woad, as well as insects and mollusks – and methods that in some cases could be traced back to Greek or even Egyptian times, were not particularly interested in innovation.

There were, however, many problems with natural-source dyes. They varied enormously in quality, and the plant-based colors changed each year depending on growing conditions. Many of the sources of dyestuffs were also produced outside Europe, increasing cost and decreasing reliable access. These problems were felt most directly in Britain, whose dominance of international trade was based in large part on the textile industry. Perkin's aniline dye came from domestic sources and was much more uniform in quality. Perkin had fortunate timing, too, because, shortly after he started production of dye, Queen Victoria chose mauve as her gown color for the marriage in 1858 of her daughter, the Princess Royal Victoria, to Prince Frederick William of Prussia. Overnight, mauve was the most desirable color in fashion. Although the colors in the Queen's outfits were not aniline, the initial reluctance of dyers to buy the new product was overcome by the demand that suddenly outstripped the supply of natural-source dyes.

If Perkin's work had been concerned only with bringing more intense, uniform, and lasting color to the world, it would have been a significant, if minor, scientific achievement. Our world is full of colors that are the chemical descendants of his work. Everything from color photography to clothing to the finish on a new automobile can be traced back to Perkin's discovery and commercial development. However, the Perkin Medal, named after and first awarded to him in 1906 by the Society of Chemical Industry (American Section), was not dedicated to him because he was a commercial success. It was created in recognition of his effect on modern chemistry, especially organic chemistry.

At a monetary level, the commercial success of aniline, and later alizarin, dyes was a dramatic demonstration that research could have real applications. A chemist could get rich by taking a discovery out of the laboratory and into the marketplace. This fact was not lost on many young scientists. A number of the world's leading chemical companies were formed because of artificial dyes. Farbenfabrik vormals Friedrich Bayer (now known simply as Bayer) was founded by Friedrich Bayer (1825–80) and J. Weskott (1821–76) in 1863 to produce fuchsine and other dyestuffs. The Aktiengesellschaft für Anilinfabrikation, better known today as AGFA, was organized by Paul Mendelssohn-Bartholdy and Carl Alexander Martius in 1867. Martius had worked for Hofmann in London at the Royal College of Chemistry; he helped to entice Hofmann to return to Germany in 1865 with the promise of more funding and new laboratories.

As economically successful as the dye industry might be, Perkin's work opened the door to organic synthesis. The tools that made aniline dye possible were the very tools that made almost all modern organic products possible. Thus, the economic impact of the exploitation of coal-tar research was enormous.

By the time Perkin retired from the dye industry at the very young age of 36, the relationship of chemical research and commercial exploitation had been completely altered. Although it would be unfair to say that this was solely the result of his discovery of mauve dye, his work was both crucial and iconic.

It is also somewhat ironic that Hofmann, who had been Perkin's teacher, could not be kept in Britain to continue to foster the next generation of chemists but moved back to Germany. By 1878 the value of coal-tar production in Britain was worth £450,000, but in Germany its value was the equivalent of over £2 million annually. Seventeen artificial-dye factories were operating in Germany compared to only six in Britain. This created a positive feedback loop. Because there were many more jobs for chemists in Germany, chemistry was an attractive educational and career path. In turn, the number of chemists extended the range of research being carried out, which led to new products and a growing demand for chemists to manage the expanding lines of chemical production. Chemical production in turn meant industrial plants that needed engineers, builders, and technicians. The new chemical processes called for new technology in steel production and parts manufacturing, which could be made available for other industries. The best and brightest chemists were guided into pure research, with the next tier filling out the ranks of instructors, and the rest being absorbed by industry. By 1897 Germany had more than 4,000 chemists in nonacademic positions, while British industry employed fewer than 1,000.

7.14 CHEMICAL DESCENDANTS AND RELATIVES OF ANILINE DYES (A BRIEF SAMPLE)

DERIVATIVES OF COAL TAR	
Dyes	Mauve (anilines), Picric, Alizarin
Pharmacology	Aspirin, Codeine, Quinine
Chemical weapons	Chlorpicrin, Brombenzylcyanide
Artificial flavours and scents	Vanillin, Coumarin
Explosives	Toluenes such as Trinitrotoluene (TNT)
Plastics	Bakelite
Biology	Methylene Blue Cell Stain

Conclusion

By the end of the century, Britain was still the most powerful nation on Earth, but its position was increasingly being challenged. In the great game of the colonial era Britain had the best colonies and controlled the seas, but Germany created a scientific and industrial powerhouse and was getting ready to use it. In the conflict that was to come, Germany turned to its scientists, especially its chemists, to overcome its disadvantages. The world of polite amateur gentleman scientists and upper-class academicians was ripped apart. Science was moving from understanding the world to mastering it, and few scientists working in the nineteenth century had any idea just how brutal scientific utility could be.

Essay Questions

1. How did the catastrophists and uniformitarians explain the origins of the Earth?

2. What was Darwin's theory of evolution by natural selection?

3. What were the roots of social Darwinism?

4. How and why did science become professionalized in the nineteenth century?

5. How did Mendeleev bring organization to chemistry?

CHAPTER TIMELINE

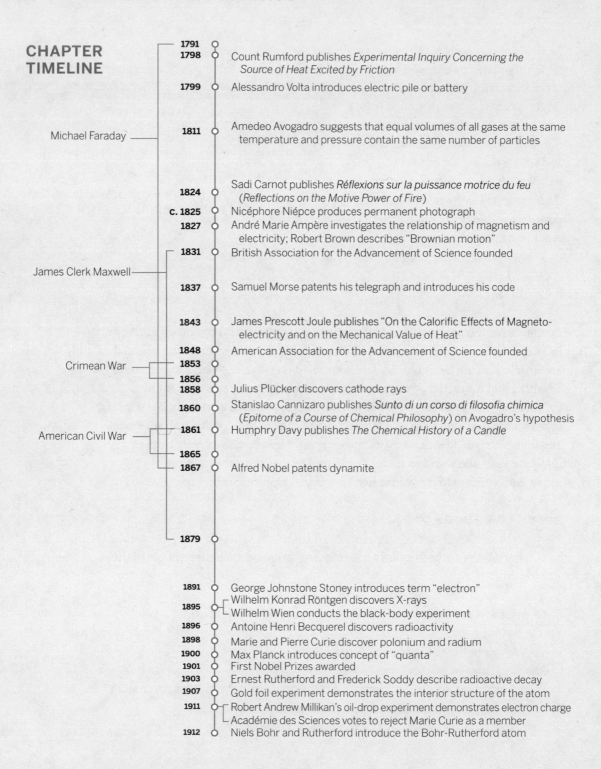

1791

1798 Count Rumford publishes *Experimental Inquiry Concerning the Source of Heat Excited by Friction*

1799 Alessandro Volta introduces electric pile or battery

Michael Faraday — **1811** Amedeo Avogadro suggests that equal volumes of all gases at the same temperature and pressure contain the same number of particles

1824 Sadi Carnot publishes *Réflexions sur la puissance motrice du feu* (*Reflections on the Motive Power of Fire*)

c. 1825 Nicéphore Niépce produces permanent photograph

1827 André Marie Ampère investigates the relationship of magnetism and electricity; Robert Brown describes "Brownian motion"

1831 British Association for the Advancement of Science founded

James Clerk Maxwell — **1837** Samuel Morse patents his telegraph and introduces his code

1843 James Prescott Joule publishes "On the Calorific Effects of Magneto-electricity and on the Mechanical Value of Heat"

1848 American Association for the Advancement of Science founded

Crimean War — **1853**
1856
1858 Julius Plücker discovers cathode rays

1860 Stanislao Cannizaro publishes *Sunto di un corso di filosofia chimica* (*Epitome of a Course of Chemical Philosophy*) on Avogadro's hypothesis

American Civil War — **1861** Humphry Davy publishes *The Chemical History of a Candle*
1865
1867 Alfred Nobel patents dynamite

1879

1891 George Johnstone Stoney introduces term "electron"

1895 Wilhelm Konrad Röntgen discovers X-rays
Wilhelm Wien conducts the black-body experiment

1896 Antoine Henri Becquerel discovers radioactivity

1898 Marie and Pierre Curie discover polonium and radium

1900 Max Planck introduces concept of "quanta"

1901 First Nobel Prizes awarded

1903 Ernest Rutherford and Frederick Soddy describe radioactive decay

1907 Gold foil experiment demonstrates the interior structure of the atom

1911 Robert Andrew Millikan's oil-drop experiment demonstrates electron charge
Académie des Sciences votes to reject Marie Curie as a member

1912 Niels Bohr and Rutherford introduce the Bohr-Rutherford atom

ENTERING THE ATOMIC AGE

8

A lthough we often associate the term "atomic" with the building of the first nuclear weapons in the middle of the twentieth century, it was actually during the nineteenth century that the atom became the focus of concentrated research. In the process, the practice of science and the relationship between science and the larger society was irrevocably changed.

In 1815 the world was Newtonian. Revolution, conquest, and war might disturb the realm of human life, but the universe swept on, a serene and inevitable clockwork governed by the laws of mass, motion, and gravity. The impact of Newtonianism stretched far beyond physics, as the Enlightenment thinkers had demonstrated, and Newton's physics itself survived over a century almost completely unaltered. Work in the physical sciences at the start of the nineteenth century tended to examine either those aspects of nature that Newton had conquered, in effect fine-tuning the Master's work, or applying Newtonian principles to those subjects he had not investigated, such as thermodynamics and electricity. The objective of most researchers was to add the new material to the cohesive system of the Newtonian worldview. As this work progressed, new tools became available, and with the new information they provided, new theoretical constructs were needed to account for the discoveries being made, thus challenging Newton's dominance. Two powerful lines of inquiry coming from these new investigations were the study of energy (characterized by a view of the universe based on waves and fields of force) and the study of matter (built on a corpuscular view of nature). These two approaches seemed to propose alternative pictures of the universe. As scientists looked ever more closely at the two subjects, what had seemed like incompatible views began to intersect, eventually coming together when Einstein's work revealed the interconnection of matter and energy.

Before that insight could be reached, however, science had to be transformed. Personal patronage was increasingly replaced by institutional funding. Science itself became more specialized, with disciplines splitting into subdisciplines; for example, chemistry was divided into organic and inorganic branches. In this era the role of the scientist became more professional, and a profusion of educational and research institutions were created, particularly in Britain, France, and Germany. The nineteenth century was also a turning point for the speed at which scientific discoveries were turned to utilitarian ends. Electricity and magnetism, for example, went from scientific objects of study in the eighteenth century to industrial applications with telegraphy and commercial electrical generation by the end of the nineteenth century.

Mastering Electricity

Although it is certain that Newton was aware of electrical phenomena, if only from reading William Gilbert, it was one of the topics untouched in his works. Investigations of electricity in the eighteenth century had been hampered by the difficulty of controlling and generating electricity. The invention in 1799 of the electric pile or battery by Alessandro Volta (1745–1827) gave scientists a consistent, controllable, and quantifiable flow of electricity and made it both an object of study and a new laboratory tool. In the spring of 1820 Hans Christian Oersted (1777–1851) put this tool to work, performing a demonstration in his home on electrical heating in a wire carrying a current. He also planned to do a demonstration on magnetism and had a compass set up nearby. During the heating experiment he noticed that the compass needle moved when current was applied to the wire. Although earlier natural philosophers such as Gilbert had seen both phenomena as similar unseen forces acting at a distance, this observation was the first experimental indication that electricity and magnetism were related. When the effect was demonstrated at the Académie des Sciences in Paris in September, it was observed by André Marie Ampère (1775–1836). Ampère had wide interests in chemistry, physics, psychology, and mathematics and had been elected to the Académie in 1814 in the mathematical section. When he saw Oersted's demonstration, he recognized its importance and turned to the study of the interaction of electricity and magnetic fields. The product was the 1827 paper "Mémoire sur la théorie mathématique des phénomènes électrodynamique uniquement déduite de l'expérience," which presented both experimental and mathematical demonstrations of electrical and magnetic behavior, including the inverse square law for magnetic action. Like gravity, the attraction of a magnet decreased by an amount proportional to the square of the distance from the magnet. Ampère's work put the study of electricity and magnetism on a solid mathematical foundation. In the same year Georg Simon Ohm (1789–1854) added resistance to the explanation. In *Die galvanische Kette, mathematisch bearbeitet (The Galvanic Circuit Investigated)*, Ohm presented his law, stating that the electromotive force was equal to the current multiplied by the total of the resistance of the circuit. In this formulation, Ohm offered a way to quantify the use of electricity by components (devices such as electromagnets and even wire) in a circuit. Ohm's original formulation was presented as:

$$x = \frac{a}{b+l'} \quad \text{which we now write as} \quad I = \frac{V}{R}$$

where I' is the current through the conductor in amperes, V is the potential difference across the conductor in volts, and R is the resistance of the conductor in ohms.

The relationship between electricity and magnetism led a number of researchers to construct electromagnets by wrapping wire around an iron core. When a current was run through the wire, the iron was polarized and produced a magnetic field. Since the wire had to be insulated from the core and from itself (or it would short circuit), and since wire did not come with an insulating layer, most early electromagnets contained few loops so that the wire would not touch itself, and varnish was used to insulate the wire from the iron core. Joseph Henry (1797–1878), working in Albany, New York, created a more powerful magnet in 1827 by covering wire with silk thread and winding 10.6 meters of this silk-covered wire around a horseshoe-shaped iron core. This was a very successful experiment, and Henry went on to create more and more powerful magnets, one of which could be used to lift over 900 kilograms. While these magnets clearly had utility, his other work on electromagnets would have more far-reaching implications. In 1831 he created an oscillating device that used two magnetic coils to attract and repel a bar pivoted like a seesaw. (See figure 8.1.) From this came the telegraph, a crucial tool for the maintenance of both national and business empires and the first step toward the electrical and electronics industries.

8.1 HENRY'S ELECTRIC MOTOR FROM *AMERICAN JOURNAL OF SCIENCE* (1831)

Henry went on to become professor of natural philosophy at the College of New Jersey (which later became Princeton University). In 1846 he became the first secretary of the newly founded Smithsonian Institution and later became president of the American Association for the Advancement of Science (1849–50) and president of the National Academy of Science (1868–78).

Henry also discovered that the relationship between magnetism and electricity worked both ways through electromagnetic induction; from this finding he created the first dynamo in 1830. By moving a wire through a magnetic field, a current was created in the wire. Although he was likely the first to create a dynamo, Henry did not publish his results, and so the scientific credit for the device went to Michael Faraday (1791–1867) instead.

Faraday's path to science was an unusual one. The son of a blacksmith, his early education was limited, but he was apprenticed to a bookbinder and read voraciously. After reading the *Encyclopaedia Britannica* article on electricity, he devoted himself to science. Faraday attended several public lectures given by Sir Humphry Davy (1778–1829) and corresponded with him. In 1812 Davy was temporarily blinded in a laboratory accident and hired Faraday as his assistant, the beginning of the younger man's scientific work in chemistry. Faraday had some notable successes as a chemist, including the discovery of what he called bicarburet of hydrogen, known today as benzene. His most popular chemical work was *The Chemical History of a Candle*, a series of six lectures for young people that was later published as a book in 1861.

Despite his success as a chemist, around 1821 he shifted his work more and more away from matter and toward forces. He was convinced that electricity, magnetism, light, heat, and chemical affinity were all aspects of the same phenomenon and that this phenomenon, rather than being based on some kind of fluid movement, was really a form of vibration. Just like a plucked violin string will cause a matching string to vibrate at a distance, Faraday reasoned that an electric "vibration" should be detectable. He took an iron ring and wrapped one coil of wire on one side to be connected to a voltaic pile and a matching coil of wire on the opposite side leading to a compass that would show the presence of a magnetic field. When the current was applied, the compass needle moved, demonstrating the principle of induction. (See figure 8.2.)

Faraday then demonstrated that the relationship between electricity and magnetism was dynamic by spinning a copper disc between the ends of a horseshoe magnet. When the disc was in motion, a current was created, but when it was at rest, no electricity was generated. It was only by passing through the field of the magnet that anything happened. In other words, the real energy of the magnet was in the space around it, not within the magnet itself, which only concentrated the forces. Faraday argued from these experiments that electricity and magnetism could be understood as a kind of strain (somewhat analogous to squeezing a spring) that affected the structure of matter. He found confirmation for this idea in an experiment in which he passed polarized light through a strong magnetic field; the plane of polarization was rotated, indicating that the light had been moved by the field. As he had thought, light was one element of the electromagnetic phenomena. Yet his explanation of fields of force was not accepted by his contemporaries, partly because in the increasingly quantitative orientation of physics, Faraday's work was not elucidated in clear mathematical terms.

That task was taken up by James Clerk Maxwell (1831–79). Maxwell was not an early prodigy like Newton, but his insight and mathematical skills led to recognition of his talents by the time he graduated in 1854 from Trinity

Iron Ring

Voltaic Pile (Battery)

Compass

8.2 FARADAY'S IRON RING

College, Cambridge, the same college Newton had attended. He went on to a series of senior academic appointments at Cambridge, ultimately becoming the first Cavendish Professor of Experimental Physics in 1871, when he was responsible for the design of the new Cavendish Laboratory. Maxwell was not the first choice for the position, but both Lord Kelvin and Lord Rayleigh (John William Strutt, 1842–1919) turned it down. The Cavendish Laboratory was named for the Cavendish family, who largely financed the original construction. Henry Cavendish had worked on a range of topics such as the nature of airs (among other things determining that water was a compound) and measuring the density of the Earth. He had also worked on electricity, but he did not publish. The Cavendish, a teaching laboratory in physics, was a key component in the development of a new course of studies known as the Natural Sciences Tripos, begun in 1851 at Cambridge to train more scientists to compete with the French and Germans. The Cavendish continues to operate today as one of the great centers of teaching and research.

Maxwell had the profound ability to create mathematical models of physical phenomena. He took Faraday's field theory and its elegant experiments and set them on a sound mathematical footing. Because he argued that a disturbance in a field was propagated through it at a particular velocity, he was able to calculate the velocity that an electromagnetic wave traveled, which was 3.1×10^{10} centimeters per second or about 300,000 kilometers per second, equal to the experimentally determined speed of light. Maxwell concluded that this could not be a coincidence. He predicted, but did not live to see it proven, that the electromagnetic spectrum would be observed at frequencies far below and far above those of light.

Maxwell's work set the course for many aspects of physics by tying several problems together. It not only linked electricity, magnetism, and light but linked them as fields and waves in fields. This undermined the Newtonian view of light as particles but neatly accorded with experimental evidence for the wave nature of light that had been building since the work of Huygens and Thomas Young. Thus, the big questions of physics seemed to be solved in a tidy fashion. Because of field theory, the universe could be viewed as a great cauldron filled with waves that propagated through a celestial fluid known as ether in which particles existed and floated around under the influence of gravity. If the system was a bit more complex than Newton had laid out, it seemed to work smoothly and was still philosophically compact; that is, the universe was governed by a small number of fixed laws that could be expressed in absolute mathematical terms.

By the time of Maxwell's death in 1879 both the scientific and the industrialized worlds were very different than at the end of the Napoleonic age. It was now

8.3 STEAM ENGINE AND DYNAMO (1907)

Source: Cayambe (Wikipedia user). Licensed under the terms of CC-BY-SA.

the era of the inventor. Steam power, telegraphy, photography, printing, and smelting all introduced new technologies and transformed the social, economic, and political lives of people around the world. The material world in the industrialized countries changed from handmade or small industry products to mass production and global distribution of consumer and industrial goods. (See figure 8.3 for a photograph of a steam engine and dynamo from 1907.)

Electricity was one of the most important factors in the transformation of European and North American society. It moved from being a curiosity for the intellectuals in the French and American salons to an industrial tool. One of its first commercial applications was the telegraph. Wheatstone and Cooke's first telegraph in 1831 moved an arrow to point to letters, although the patent and their priority were challenged. In 1833 the mathematician Karl Gauss (1777–1855) and Wilhelm Weber (1804–91) built a model telegraph system that sent a signal over a distance of two kilometers. Telegraphy was made commercially viable when the first commercial electrical generator was marketed in 1834. In 1837 Samuel Morse (1791–1872) patented his version of the telegraph and his code system of dots and dashes. In 1844 Washington was linked to Baltimore by telegraph; London and Paris were linked in 1854; and the first Atlantic cable was laid in 1858. Telegraph lines followed the railways across the globe, and New York was connected to San Francisco in 1861. When the American Civil War began, news of the horrific events were circulated wherever the telegraph went. Telegrams sent by reporters were printed by newspapers using steam-driven presses and offered coverage of events around the world in hours rather than days or weeks.

The Nature of Electricity and the Science of Thermodynamics

The speed of development of electrical devices by inventors all over the world was astounding, but even as electricity was being turned into a commercial product, its nature continued to be a mystery. Faraday and Maxwell had treated electricity as a vibration or field of force, and the generation of electricity by the interaction of conductive wire and magnetic fields corroborated this view. On the other hand, the generation of electricity from chemical sources (as in batteries) suggested that there was a material basis to electricity. Chemists were concerned about electrolysis (including chemical reactions via application of electrical currents), and that led to theories about electropositive and electronegative activity. For example, positive sodium reacted with negative chlorine to form sodium chloride. Although ideas about electrolysis helped scientists to understand new aspects of atomic and molecular interaction and focused the zone of electrical activity, it did not solve the underlying questions about the origin and nature of electricity.

Part of the solution was far from direct and was one of the most debated hypotheses in modern science, being rejected and revived several times. In 1811 Amedeo Avogadro (1776–1856) proposed in the *Journal de physique* the hypothesis that equal volumes of all gases at the same temperature and pressure contained the same number of particles. This was a resolution of problems presented by Joseph Gay-Lussac's (1778–1850) work on laws of gases and by Dalton's work concerning the combination of gases. Scientists knew that one volume of oxygen combined with two volumes of hydrogen to produce two volumes of water vapor. Avogadro reasoned that this could not be unless the oxygen was actually a molecule that broke up into two atoms (or "half-molecules" as he called them) producing two water molecules for every original oxygen molecule (what is now often called "diatomic" oxygen or O_2). In its simplest form, Avogadro's hypothesis posited a fixed number of atoms or molecules in a special quantity. This quantity is now referred to as a "mole" and is the amount of a substance with a mass in grams equal to its atomic or molecular weight. Determining this basic bit of information was essential to understanding the composition of materials, but it was also a key to understanding how materials formed. For the material world to function consistently, each molecule of a particular substance had to be formed of the same elements in the same proportions, but it was impossible to understand how the elements linked together to make molecules if the precise number of each element in a molecule was unknown. Figuring out the proportions provided a tool for figuring out why some elements such as oxygen and carbon combined to make many compounds, while others such as gold and silver combined with very few other elements to form molecules.

Avogadro's hypothesis was not favorably received, because it depended on particles of the same kind being attracted to each other, which ran against the affinity theory of compounds based on the attraction of dissimilar particles (like the North and South poles of magnets attracting, while North to North repelled). Ampère attempted to revive Avogadro's idea in 1814, as did Auguste Laurent and Charles Frédéric Gerhardt (1816–56) in organic chemistry in the 1840s, but it was not until 1860 when Stanislao Cannizzaro circulated a pamphlet, *Sunto di un corso di filosofia chimica* (*Epitome of a Course of Chemical Philosophy*), at the Karlsruhe Congress that Avogadro's hypothesis began to affect thinking about the solution to the problem of determining and comparing molecular and atomic weight.[1] The Karlsruhe Congress was the first international chemistry conference, and it was attended by some of the most important scientists of the day. The meeting had been called to try to develop international standards for nomenclature and to do something about the problem of atomic weights. Avogadro's hypothesis offered a way to deal with the problem of atomic weights that everyone would eventually accept.

What at first seemed to be a matter problem became linked to electricity because matter can have an electrical charge and that charge appears to have something to do with keeping different elements together in compounds.

....

1. The term "weight" is actually incorrect, since what was being measured was the mass of the particle, but by convention molecular and atomic mass continues to be molecular and atomic weight.

Avogadro's hypothesis was much better understood by the time Svante August Arrhenius (1859–1927) arrived in Stockholm in 1881 to work on solutions and electrolytes under Eric Edlund (1818–88). According to Arrhenius, when a current was passed through molten sodium chloride (NaCl), the molecule broke apart or dissociated. Its parts were not atoms, but what he called ions; the sodium had a positive charge (Na^+) and the chlorine a negative charge (Cl^-). These ions migrated to the electrical poles, sodium to the cathode and chlorine to the anode, where they lost their charge: the result was atomic (or elemental) sodium and atomic chlorine.

This suggested that atoms and groups of atoms might themselves carry an electrical charge rather than merely being affected by electrical activity. This raised as many questions as it answered, because it seemed to require the combination of two incompatible objects – Daltonian atoms and Maxwellian electric waves. One solution was to picture electricity as a kind of particle, or indivisible unit. The Irish physicist George Johnstone Stoney (1826–1911) calculated the magnitude of what he initially called the "atom of electricity," and in 1891 he proposed the term "electron" for this unit of electrical charge.

While some scientists were picturing electricity as a kind of particle, others tried to solve the problem by recreating atoms. Cartesian particles, Newtonian corpuscles, or Daltonian atoms were all small, discrete, and indivisible particles that interacted in specific, if somewhat unclear, ways. While this was a useful concept for those studying gross matter, it presented problems for those studying magnetism, light waves, heat, and so on. A different model of the atom based on a vortex was developed by William Macquorn Rankine (1820–72) in 1849 and also by Lord Kelvin in 1867. Kelvin argued that the vortex atom was perfectly elastic and that a number of important conditions, such as thermal expansion and spectral lines, could be derived from this kinetic model.

Kinetic Theory of Gases and Thermodynamics

These new ideas originated in work on heat. Heat, or thermodynamics as it was later called, was another one of the areas that Newton had not addressed. In the Aristotelian system heat was an element. Through the eighteenth-century heat was considered to be the essence of fire or phlogiston, until Lavoisier disproved the phlogiston theory with his careful experiments. With the help of Laplace, he introduced caloric, which was an "imponderable fluid." The idea of caloric did help to explain the apparent movement of heat and the difference between heat and temperature. Caloric was thought to flow from warm areas (full of caloric) to cool areas (empty of caloric). Heat represented the total volume of caloric in a given body, while temperature measured the concentration of caloric, so that, for example, a lake could have a cooler temperature than a boiling kettle, but it contained far more caloric.

As ingenious as the idea of caloric was, it did not accord well with Newtonian physics, and, by creating a new class of matter, it also went against the trend in the nineteenth century to locate physical properties within objects rather than as additional types of things that affected objects. As early as 1738 Daniel Bernoulli (1700–82) of the famous Swiss mathematics and physics family had presented a theory of pressure based on Newtonian principles about the movement of atoms. The faster the particles moved, the greater the pressure. The theory was seen as improbable and was mostly ignored. That heat might have something to do with the state of atoms had to be demonstrated another way before it gained attention. The attack on caloric came from Benjamin Thompson, Count Rumford (1753–1814), a man with a wildly varied career that included spying for Britain, service to the Elector of Bavaria (who created him Count Rumford of the Holy Roman Empire), building a better fireplace, marrying Lavoisier's widow (the marriage lasted only one year), and overseeing the production of cannons. Rumford had initially supported the caloric theory of heat and in fact had added "frigorific radiation" (cold rays) to the theory, but it was his observation of cannon-boring equipment that led him to reject caloric. He observed that a dull drill bit, although spinning indefinitely, could not cut through the iron of a cannon but would continue to produce heat. If caloric was a substance, it should eventually be emptied out of the iron, but it was not. In 1798 he published *Experimental Inquiry Concerning the Source of Heat Excited by Friction*. It became a classic paper in physics. Caloric theory, although proven to be wrong, had led to the quantification of heat, and Rumford's work advanced the study of heat by directly linking work (kinetic action) to heat.

In 1824 Nicolas Léonard Sadi Carnot (1796–1832) produced his *Réflexions sur la puissance motrice du feu* (*Reflections on the Motive Power of Fire*), in which he scientifically analyzed the theory of heat engines or devices in which work was done by the movement of heat from one area to another, such as the heating and condensing of steam in a steam engine. Although based on the caloric theory of heat, he demonstrated that work from heat was analogous to water falling from a high point to a lower point. From this, it became possible to calculate the efficiency of heat engines, which was remarkably low. In a steam engine that cooled steam by 100° centigrade (that is, the steam "fell" from 150° to 50° as it passed through the pistons and condensed as water), the efficiency was only 0.236, which means that less than a quarter of the heat could be converted into work. Carnot's work was brilliant, but he died of cholera at the age of 36, and his work was not widely known until revived by Joule, Kelvin, and Clausius.

The relationship between work and heat was formalized by James Prescott Joule (1818–89) in a paper he read before the BAAS in 1843. Entitled "On the Calorific Effects of Magneto-electricity and on the Mechanical Value of Heat," it offered the first quantification of the mechanical equivalent of heat corresponding to a rise in temperature. He argued that to raise the temperature of 454 grams of water by one degree Fahrenheit required the expenditure of 255 meters/454 grams of work. In other words, the amount of work necessary to move a 454 gram weight 255 meters would, if converted into heat, raise the temperature of one pound of water by one degree. The paper met with silence from the scientific community, but he persevered and refined his experiments. In 1845 he presented "On the Mechanical Equivalent of Heat" to the BAAS, with his new calculation of 250 meters/454 grams. By using a paddlewheel (which produced friction as it moved through a fixed volume of water) driven by a falling weight, he refined his figures in 1850 to 236 meters/454 grams. (See figure 8.4.) By this time, his work was receiving more favorable attention.

Joule went on to work with Lord Kelvin on the study of heat and helped develop the kinetic theory of gases. That theory would link heat and pressure to the motion of particles in a gas with the claim that the average speed of the particles related directly to the temperature of the gas. Independently, Rudolf Clausius (1822–88) arrived at the same conclusions about the motion of atoms. He also introduced the term "entropy" in 1865 to describe the dissipation of energy, which always increased over time. With Clausius's work, the kinetic theory of gases – that is, the relationship between the motion of the particles and the heat of the system – was established. Heat was not a mysterious fluid but the motion of atoms and molecules. Further, the motions of all those atoms moving around could be measured, at least statistically, and that meant heat could be used as a key tool for explaining states of matter.

8.4 JOULE'S DIAGRAM OF THE MECHANICAL EQUIVALENT TO HEAT

From "A New Theory of Heat," *Harpers New Monthly Magazine* 39 (1869): 327.

The kinetic theory of gases was based on the existence of atoms and molecules, so it became an integral part of matter theory. In 1865, Joseph Loschmidt (1821–95) applied the theory to calculate the diameter of an atom and used this value to determine the number of molecules in a unit volume of gas. Clausius envisioned the "energy" of a molecule partitioned into contributions from its translational motion (movement from place to place), vibrational motion, and rotational motion. Based on this, he showed that the heat capacities of hydrogen, oxygen, and other elements in the gaseous state were consistent with the theory that their molecules were diatomic, lending support to the vapor density method for determining atomic weights.

The modern presentation of all this complex work can be encapsulated in two general laws that emerged in the 1840s and 1850s. The first law defined energy as a new fundamental concept in the physical sciences. It states that heat is a form of energy and that, in any closed system, the total amount of energy is constant. Thus, it would be impossible to create a machine in the real world that worked continually without an external source of energy, since the energy used to work the machine would be dissipated (by friction, for example) and thus become unavailable to run the machine. Another way of saying this is that you can't get more energy out of a system than you put into it.[2]

The second law of thermodynamics, the entropy law, says that in any physico-chemical process, it is impossible to convert all the energy into work. Some energy is always converted into heat and thus is not available for work. Further, in any closed system, heat will transfer only in one direction, from a warmer to a cooler region. Over time, entropy will make everything the same temperature if there is no external energy source. This is why a thermos bottle both works and doesn't work perfectly. The thermos, by reducing the rate of transfer of heat from the hot interior to the cooler exterior keeps the hot beverage warm longer, but because the barrier is not perfect (nor can it be), the contents of the thermos will eventually cool to the temperature of the surrounding world. Whether it is coffee in a thermos, or the whole universe, entropy applies. It is just a matter of time.

This idea had wider implications than just physics. Kelvin used this theory to estimate the age of the Earth based on the rate of cooling seen in a heated sphere of nickel-iron. This suggested an Earth too young for Darwin's theory of evolution and became a major objection to Darwinian evolution.

....

2. An example of this problem is trying to cool your kitchen by leaving the refrigerator door open. Since the refrigerator works by taking the heat out of the interior of the refrigerator and radiating it to the surrounding air of the kitchen, you would heat the air you were trying to cool down. And because the motor that runs the refrigerator would generate more heat from the friction of the moving parts, you would gain even more heat in the room.

Physical Chemistry, Cathode Rays, and X-Rays

Although the kinetic theory of gases bolstered belief in atoms by relating heat to the motion of particles, some scientists were coming to believe that energy – not matter – was the genuine physical reality and the appropriate basis of physical science. An increasing number of scientists argued that atoms did not exist as material objects at all. In addition to Kelvin's vortex model, Maxwell claimed that the actual existence of atoms need not be assumed for mathematical models to work, and Willard Gibbs (1839–1903) worked out his theory of thermodynamics without particles. One of the most outspoken opponents of the idea of atoms was Ernst Mach (1838–1916), who argued not only that atoms as described by atomism were hypothetical and need not exist but that theorizing about them might be misleading. Mach was one of the leading proponents of positivism, a philosophy that argued that the only reliable form of knowledge was scientific, arising from observable or empirical evidence. Since atoms could not be observed and were unnecessary for many aspects of physics, they fell into the category of metaphysics, a branch of philosophy that speculated about the origin and purpose of nature rather than looking at how nature functioned. According to Mach, nothing in science could be based on such speculation.

The anti-materialist idea even had supporters among chemists, with the influential physical chemist Wilhelm Ostwald (1853–1932) in Germany and such notable French chemists as Marcellin Berthelot (1827–1907) and Henri Le Chatelier (1850–1936) also opposed to atomism. They supported the Energetik system, according to which matter had to form a continuum, as all "matter" had to be part of the spectrum of energy. Energetik theory was based on thermodynamics and reinforced by the wave theory of light.

Many supporters of the Energetik theory called themselves physical chemists. In 1887 Ostwald, along with Arrhenius and Jacobus Henricus van 't Hoff (1852–1911), founded the *Zeitschrift für physikalische chemie* (*Journal of Physical Chemistry*), which became the leading journal in this new field. Chemistry and physics were beginning to separate both institutionally and in terms of areas of research. Physical chemistry bridged the gap by attempting to use the concepts of thermodynamics and energy to account for chemical reactivities and various physical behaviors of substances. Although physical chemistry provided some profound insights into the structure of matter and the role of forces, it was not embraced by the whole chemistry community. Some chemists objected because it depended so heavily on quantitative and mathematical analysis that seemed either superfluous or uncertain. Others, particularly those working in organic chemistry, found that the questions asked by physical chemists were simply unnecessary for their work.

The relationship between matter and energy was creating new puzzles. For example, water broke down into oxygen and hydrogen when a current was passed through it, but it was also known that these two elements would dissociate at very high temperatures. Electricity passing through a solid seemed to have no effect on the conductor, while it caused decomposition in liquids. Experiments that passed a current through gases at normal pressure either did not work or required very high voltage, which then resulted in violent sparking or even explosions. If matter was an illusion, then physicists would have to explain how energy, fields of force, and waves interacted to give what seemed to be a material universe. If matter and energy were discrete, how was an electrical charge generated from matter? What was needed to resolve these issues was a new way to study the component parts, and that meant new laboratory tools.

At very low pressure electricity would pass through gases, but it was not until the glassblower Heinrich Geissler (1814–79) devised a method of producing a good vacuum tube that research could be conducted on the phenomenon. In 1858 Julius Plücker (1801–68) noticed that when a current was passed between electrodes inside a vacuum tube, a greenish glow appeared on the cathode (negative terminal). This glow was the same regardless of the metal used for the terminal, so he concluded it was an electrical phenomenon rather than a property of the material of the electrode. Plücker also demonstrated that the luminescence was affected by a magnet placed near it.

Plücker was followed by Johann Wilhelm Hittorf (1824–1914) in Germany and William Crookes (1823–1919) in Britain. In 1869 Hittorf (who had been Plücker's student) demonstrated that whatever was being projected through the vacuum traveled in straight lines and could cast a shadow if there was an obstacle in its path. Hittorf's findings were confirmed by Eugen Goldstein (1850–1930), who named the radiation *Kathodenstrahlen* or cathode rays. Crookes independently found the same results. He also attempted to see if the rays exerted a force by placing a paddlewheel inside the tube. The cathode rays rolled the paddlewheel away from the cathode and toward the anode. From this he argued that the rays were really particles with a definite mass, although it was later shown that it was, in fact, the trace gas molecules that caused the paddlewheel to rotate.

Work with cathode rays did not resolve the problem of atomic composition, however. For physical chemists (particularly in Germany), the behavior of the rays seemed to confirm the wave nature of the universe, while in Britain Crookes's work was seen as confirming the particle nature of matter.

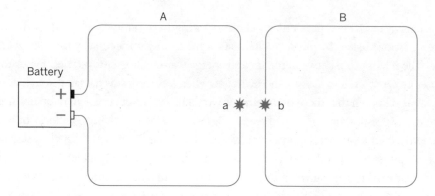

8.5 HERTZ'S SPARK GAP EXPERIMENT

In Germany, Heinrich Rudolf Hertz (1857–94) continued to explore Maxwell's reasoning about waves. He created an elegant experiment to demonstrate that electromagnetic radiation could be generated, broadcast, and detected at a distance by making electricity jump a small gap in the air, thus producing an electric spark. To detect the waves, he used a rectangle of wire with the two ends almost touching. If electromagnetic radiation created by the original spark passed over the rectangle of receiving wire, it should produce a current and cause a spark to cross the small gap, which it did. (See figure 8.5.) These "Hertzian waves," or radio waves as they became known, validated Maxwell's ideas and extended both the spectrum of electromagnetic radiation and the ability to examine the phenomenon experimentally.

The radiation spectrum was further expanded by the discovery of X-rays by Wilhelm Konrad Röntgen (1845–1923). In 1895 Röntgen, investigating ultraviolet light, used a cathode ray tube and paper coated with barium platinocyanide that fluoresced if exposed to ultraviolet. When he covered the discharge tube in black paper, the coated paper still glowed. Even when placed in an adjoining room, the coated paper glowed when the current ran in the cathode tube. The rays seemed to originate from the glass at the end of the tube where the cathode rays struck. A range of materials were examined to determine the action of the mysterious rays. The denser the material, the less penetration was found. When Röntgen exposed photographic plates to the rays, he discovered that they recorded images. One of the first "röntgenograms" was of his wife's hand and clearly showed the bones and a faint outline of the flesh. The potential medical use for the discovery was immediately apparent, and the first medical use of X-rays took place only a few months after their discovery.

Röntgen's work and the medical application of X-rays would not have been possible without the invention of photography. Nicéphore Niépce (1765–1833) is credited with creating the first permanent photograph around 1825. His system was improved by Louis Daguerre (1787–1851) and others. By 1884 George Eastman (1854–1932) had introduced film photography, which remained the foundation of modern photography until the introduction of digital imaging. Photography is perhaps the best example of a commercial invention that was then brought into the laboratory, where it became a tool for everything from astronomy to cell biology. The use of photography raises an interesting philosophical issue for investigators about the limits of experimental reliability and objectivity, since they did not have access to the chemical composition of the film (a trade secret) and therefore could not control all the variables for their experiments.

Later work by Max von Laue (1879–1960) demonstrated that X-rays had the same fundamental electromagnetic properties as light, but at a much higher frequency, and that crystals would diffract X-rays and produce a consistent pattern. This idea was then used by William Henry Bragg (1862–1942) and his son William Lawrence Bragg (1890–1971) to create the study of crystallography, which mapped the interior structure of crystals by the analysis of X-ray diffraction images. This technique had major consequences in a wide range of fields, particularly as the clue to the structure of DNA. The other development from von Laue's work was the creation by Karl Manne Georg Siegbahn (1886–1978) of X-ray spectroscopy, which allowed a better understanding of the elements and a method of determining the composition of materials by the examination of the electromagnetic spectrum emitted by matter when heated. If a substance was heated, it produced a consistent and characteristic pattern of light. This was useful for identifying elements and would even be used to determine the composition of stars.

8.6 RÖNTGEN'S X-RAY OF ALFRED VON KÖLLIKER'S HAND, 1896

Röntgen was showered with awards for his discovery of X-rays, including the Nobel Prize in Physics in 1901, the first year the prizes were awarded. Named after Alfred Bernhard Nobel (1833–96) and his family, the prizes became the most prestigious award in science. Nobel was a self-taught scientist, who was particularly interested in the branch of chemistry dealing with explosives. The foundation of the Nobel fortune was dynamite, patented in 1867, and a range of other explosive products including smokeless powder. Industrial explosives were in high demand in the age of big engineering projects such as canal building, rail construction, and large-scale mining, but they were extremely dangerous. Nobel's products were stable and far more predictable. Military applications for explosives were also on the rise as European military forces expanded rapidly in the age of colonial empires.

Nobel left much of his vast fortune from the explosives industry for the creation of a series of prizes in science, peace, and literature. He wanted to reward those whose work conferred the "greatest benefit on mankind." People have speculated that he created the prizes in response to the criticism that his fortune was gained by death and destruction. The Nobel Prizes in physics, chemistry, physiology or medicine, peace, and literature were to be awarded annually and selected by special committees. The Swedish Academy oversaw the literature prize, while the Royal Swedish Academy of Sciences was responsible for physics, chemistry, and physiology/medicine. The peace prize was selected by a committee established by the Norwegian Parliament.[3] Because the prize system required the coordination of several academies and two governments, it took several years after Nobel's death in 1895 to organize.

Although some have argued that the Nobel Prizes are not always given to the best candidates, the science prizes have generally been awarded to people whose discoveries or body of work have had a major effect on the discipline and often on the whole field of science. They also represent a conduit for public exposure to science, since the awards were, and are, a major news event. Winning Nobel Prizes has become a matter of national pride. In addition to the prestige and a gold medal, the winner also receives a substantial cash prize. The amount of the prize varied from year to year based on the returns on the Nobel endowment. In 2001, the cash value was set to 10 million kronor (worth about $1.17 million US in 2021 dollars). It was reduced from 2012 to 2019 but returned to 10 million kronor in 2020.

----

3. It is not clear why Nobel gave the responsibility for the peace prize to the Norwegians, although they were (and continue to be) active in the international peace and disarmament movement. The economics prize was created in 1968 by the Bank of Sweden and was not part of Nobel's original plan.

Resolving the Wave/Particle Dilemma

While X-rays were making news, the battle to understand just what was going on with particles and the electromagnetic spectrum continued. In an attempt to better understand the strange particles of the cathode tube, Hertz had shown, by constructing a cathode tube that sent a beam of rays between two electrically charged metal plates, that cathode rays were not deflected from their course when they passed through an electric field. J.J. Thomson (1856–1940) thought that Hertz's conclusion had to be wrong and proved that the rays *were* deflected when passing through an electrical field. He did this by creating a better vacuum and removing gas particles that had interfered with the electrical charge on the cathode rays. The discovery had many consequences: it raised further questions about the nature of the rays, and it also formed the technical foundation for modern consumer electronics as the principle behind cathode-ray tube televisions and computer monitors.

Although J.J. Thomson seemed to have transformed Hertz's ray into a particle, it was a strange particle that he found. Its path could be deflected by both a magnetic field and an electrical field. The path of the deflection suggested a negative charge on the particle, unlike X-rays which were unaffected by electrical fields. Because the velocity and path of the cathode ray could be controlled by magnetic and electrical fields, and the strength of those fields was known, it was possible to estimate the mass of the strange particle by calculating its ratio of mass to charge. The electron's mass was 1/1836 that of a hydrogen ion, or in modern terms 9.1091×10^{-28} grams. This particle appeared to be Stoney's electron.

While knowing the mass of the electron was a breakthrough, it did not automatically determine the charge of the particle. This was done experimentally in 1911 by Robert Andrew Millikan (1868–1953), who used droplets of oil floating between two oppositely charged plates to measure the electron's charge. By observing the motion of a tiny oil droplet that absorbed ions from the air (he used X-rays to ionize the atmosphere), he could calculate the charge. He reasoned that the charge could have only one size and that was the unit charge of the electron, which was the same for all electrons in all atoms.

While the mass and charge were being figured out, there remained a troubling question: Where did the particle go after it hit the end of the cathode tube? The fluorescence associated with cathode rays linked the particle to light, but that just brought the question around full circle. Was light a particle or a wave?

Most discoveries about the electromagnetic spectrum seemed to confirm the wave nature of the universe, but there were problems with the wave theory as well. Waves, by definition, had to be propagated through a medium. Just as there is no sound to be heard from a bell rung in a vacuum, if light were a wave it had to have a medium to travel through, or there would be no electromagnetic spectrum to observe. The characteristics of the electromagnetic medium, or the ether as it was called, would have to be very particular, and for some scientists the electromagnetic ether seemed like a throwback to the Cartesian idea of vortex and plenum.

If the supporters of the Energetik theory were right, that all "matter" was simply a manifestation of waves, there remained the question of how waves could manifest themselves as a solid, since waves had the property of passing through each other. In other words, why couldn't a person walk through a wall, if both the person and the wall were not material but composed of waves? Two lines of work not only led to a resolution of the wave/particle problem but also, ironically, unraveled the very foundation of the Newtonian system. The first line came from the discovery of radioactivity and the second from a logical conundrum that was created by the impossibility of both Newton's system of mechanics and wave theory being correct.

In 1895 Antoine Henri Becquerel (1852–1908) began a detailed study of fluorescence and phosphorescence, an area he had been interested in for some time, but which was made more interesting by the work of Röntgen on X-rays. Becquerel wondered if fluorescing materials could in turn produce X-rays or cathode rays. He chose to work with uranium salts because they were known to fluoresce strongly. First, he exposed the crystals to strong sunlight, and then placed them on a sealed photographic plate. When the plate was developed, there was a darkening where the crystals had been, indicating the passage of rays from the uranium salts through the covering paper.

On February 26 and 27, 1896, overcast skies prevented Becquerel from continuing his study, so he left the crystals and photographic plate in a light-proof drawer. Curious to see if there was any residual trace of the rays from the crystals, he developed the plate anyway and discovered that the dark image was just as strong as the image from the salts exposed to light. Whatever had darkened the plate came from the sample, not from something absorbed and later emitted. Closer examination of the uranium showed that the radiation could not be reflected like ordinary light but could affect the electrical charge of objects on which it fell.

At about the same time, Marie Sklodowska Curie (1867–1934) began her work on radioactivity. She used a piezo-electric quartz electrometer that had been invented by her husband Pierre Curie (1859–1906) and his brother Jacques (1856–1914). This tool could measure very small levels of electrical charge, and she used it to identify substances that demonstrated the effects noted by Becquerel. Only uranium and thorium gave off the ionizing radiation, but when she tested a sample of pitchblende, she found that its ionizing power was greater than pure uranium. Pitchblende was for the Curies something like coal tar had been for organic chemists. It was composed largely of an oxide of uranium, U_3O_8, but also contained other components in very small quantities. Uranium was a rare material whose main use was as a coloring agent for glass-making, where it produced a beautiful blue glass. After months of work refining the raw ore or pitchblende, the Curies isolated a new radioactive element that they named *polonium*, after Poland, Marie's homeland. Further refining work produced a second radioactive element, *radium*, which they announced in December 1898. They had processed eight tons of pitchblende in order to produce one gram of radium compound. It is likely that the long-term exposure to radioactive substances led to Marie Curie's death from cancer in 1934.

Becquerel and the Curies were awarded the Nobel Prize for Physics for their work on radioactive substances. Tragedy befell when Pierre was killed in a traffic accident in 1906, but Marie continued on with her work. One of the few women allowed into the male-dominated world of science, she was the first female Nobel winner and the first female professor at the Sorbonne. In 1911 her name was put forward for membership in the Académie des Sciences, but in a highly public and acrimonious fight, she was rejected. Many still believed that women were unsuited for scientific work or that she was riding on her husband's coattails. Not only misogyny but other factors affected the rejection – anti-Semitism, the split in France between conservatives and liberals, and nationalism. Although Marie Curie was Catholic, her name suggested Jewish heritage, and she was caught in the animosity arising from the Dreyfus Affair.[4] Because she was Polish, she also became the target for radical nationalists, and, as a liberal, she was seen as a threat by the conservatives both inside and outside the Académie. When she won a

---

4. Alfred Dreyfus, a Jewish officer in the French Army, was falsely accused of sending secret information to the Germans and was imprisoned for life. A protect, spearheaded by the novelist Émile Zola, led to a retrial and an indictment of the army's practices and verdict. Dreyfus accepted a pardon to end his torment. The Affair caused deep fissures in French society and politics.

second Nobel Prize that same year in chemistry, her opponents decried it as a political gesture. Marie Curie remained the only person to receive two Nobel Prizes until 1962 when Linus Pauling won his second (Chemistry 1954, Peace 1962).

Despite the detractors, Marie Curie's work not only was instrumental in understanding radioactivity and discovering new elements but also opened up understanding of the very structure of matter. Becquerel and others recognized that the ionizing radiation was not of a single type. Some radiation could penetrate only a thin layer of metal foil, while some penetrated much deeper. In addition, some of the radiation from radioactive material could be deflected by an electrical field. These different characteristics were identified as different types of rays, labeled alpha and beta rays around 1900, with gamma rays, which had more penetrating power, being identified and named around 1903.

The Atom Deciphered

While the hottest topics in physics seemed to be X-rays, radiation, and radio waves, there was a quiet experimental program studying colloids as a method of demonstrating the physical reality of atoms and matter. There were two reasons why this line of research was overshadowed by other discoveries. The first was that from the beginning colloids were associated with completely mundane things such as glue, paint, and a host of industrial and household products. The second reason was that colloids were more closely associated with organic and biological studies than with the fundamental nature of matter.

Around 1900 the prevailing theory about cell material in plants and animals was based on colloid chemistry. Colloids were identified by Thomas Graham (1805–69), a Scottish chemist and the first president of the Chemical Society of London. He classified a group of materials that would not diffuse through a membrane as colloids (as opposed to crystalloids, which did pass through the membrane). He thought these substances – like mastic, fats, paints, and parts of blood – had certain unique properties such as the ability to form gels. Cell chemistry was based on a belief that these colloids were largely undifferentiated masses with the chemically active parts – the enzymes that controlled cell activity – embedded or attached to the mass. This was called the "träger theory" from the German word for "carrier."

It became significantly easier to study colloids when Richard Zsigmondy (1865–1929) and his assistant Henry Siedentopf (1872–1940) constructed the first ultramicroscope in 1903. The ultramicroscope was a marvel of optical engineering and effectively reached the limit of optical observation. It allowed the examination of particles down to 5 millimicrons in diameter (although its general operating range was 20 to 200 millimicrons) by illuminating the object from the side against a dark background, just as one might see floating dust motes in a beam of sunlight. This allowed a quantitative evaluation of colloidal material. (See figure 8.7.)

8.7 ULTRAMICROSCOPE

A further step was taken when Theodor ("The") Svedberg (1884–1971) developed the ultracentrifuge. This device, which spun samples of particles in a fluid medium at up to 100,000 gravities, separated materials too small to be analyzed using other methods. Svedberg used the ultracentrifuge to study organic colloids such as hemoglobin and made a surprising discovery. Rather than finding a range of particle sizes in a sample of hemoglobin (and other such substances), he found that all the particles were the same size. This suggested that rather than being undifferentiated and random clumps of organic material, the chemical components of cells were uniform. These findings were hard to accept at first, because biochemists had trouble believing that a single molecule could be as large as the tests suggested. Hemoglobin, for example, had an atomic weight of 68,000. Biochemists had expected cellular material to be composed of small components and initially argued that the highest molecular weight for a single functional molecule could not be greater than 16,000. The hemoglobin result seemed almost absurd to some scientists, and Svedberg repeated the experiment several times to confirm the results.

Into this debate came Hermann Staudinger (1881–1965). Staudinger had a rebellious streak, having opposed German chemical warfare during World War I, and he was prepared to challenge fellow chemists regardless of their status. He was, therefore, willing to oppose the accepted opinion and champion the idea of large organic molecules. In 1924 he proposed the name "macromolecule" for the large aggregates that were showing up more and more in organic chemistry. He was roundly criticized for suggesting that these colloid compounds could be single functional molecules, and some chemists even suggested poor laboratory practice had led to the error. In 1929, as more work was done on colloids, Staudinger went to the Notgemeinschaft der Deutsche Wissenschaft (the German Research Foundation) and asked for money to purchase an ultracentrifuge. When his request was refused, he turned to other methods to determine the molecular weight of polymers, or long-chain molecules that were constructed from smaller units. He found that viscosity was related to molecular weight and that, by measuring viscosity, he could estimate the weight and size of molecular chains. With simple tools, he predicted a number of molecular weights that were confirmed by other methods. He also chemically altered a macromolecule without changing its molecular weight. While this was not the *experimentum crucis* that changed chemistry in a single moment, it did undermine the colloid theory of organic substances as being indeterminate masses. For the geneticists, the path to the ultimate control system of the cell became much clearer when macromolecules were accepted and their composition laid open to study. As long as organic molecules were thought to be indeterminate or even random aggregations, there was little way to organize or systematically study enzymes, hormones, and, ultimately, DNA.

In 1895 Jean-Baptiste Perrin (1870–1942) used the study of colloids to demonstrate that cathode rays were not waves but charged particles. He did this by showing that the rays could transfer a measurable charge to another object. Flying in the face of the popular Energetik theory, he set out to demonstrate the physical nature of atoms and molecules. Starting around 1908 he created and painstakingly observed, using the ultramicroscope, the behavior of tiny colloid particles moving through water in a flask. The particles were too small to settle to the bottom of the container and, like dust motes dancing in air, seemed to bounce and zigzag. This random motion was called Brownian motion, since it had been first noted by the botanist Robert Brown (1773–1858) in 1827. Perrin argued that the changes in direction were due to collisions of the particles with molecules of the liquid. Since he knew the mass of the particles he had made, he could calculate the mass and kinetic energy of the water molecules. From that, he could calculate the number of molecules in a given volume of liquid and provide an experimental confirmation of Avogadro's hypothesis. From the hypothesis comes a method for calculating the specific number of particles in a given mass. This number is known as Avogadro's number and is equal to 6.022×10^{23} particles in a mole. In other words, one mole of hydrogen has the same number of atoms as one mole of gold or one mole of hemoglobin molecules. With Avogadro's number, one can figure out how many atoms exist in a given mass of an element or compound.

Perrin published his conclusions in 1909, apparently unaware that his experimental work neatly confirmed the theoretical argument for the physical existence of atoms and molecules that had been published in 1905 by Albert Einstein (see Chapter 9).

While the work of Perrin and Einstein supported the actual existence of atoms and molecules, there was still the question of what the structure of the atom looked like. Around this time, the general conception of the atom was the "plum pudding" or "raisin bun" model, which pictured the atom as a single positive mass, with negative charges (the "raisins") distributed throughout. The model was reminiscent of the "träger theory." The model was not very elegant, but it fit a number of the requirements that had to exist because of the chemical and physical behavior of the atom, such as the ability to gain and lose electrical charges and to form bonds. (See figure 8.8.)

Radioactivity disturbed this picture, since radioactive substances seemed to be shooting out bits of the atom. In 1903 Ernest Rutherford (1871–1937) and Frederick Soddy (1877–1956) published an article arguing that radioactive substances were in fact going through a series of transformations rather like a china figurine falling down a set of steps; the pieces that broke off were the alpha and beta rays. This has sometimes been called "modern alchemy" since it was an actual transmutation of matter, but unlike the medieval alchemy, it took rare metals and turned them into base metals – uranium into lead!

By suggesting that atoms were not permanent structures, Rutherford and Soddy undermined the certainty both of the corpuscular theory that had been the foundation of matter theory since the Greeks and of Newtonian corpuscularianism. In fact, the discovery that cathode rays, radioactivity, and X-rays had a material foundation suggested that atoms had sub-parts that existed separately

Positively Charged Material

The Raisin Bun Model

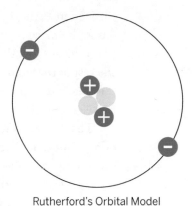

Rutherford's Orbital Model

8.8 RAISIN BUN MODEL AND RUTHERFORD'S ORBITAL MODEL

from the whole. To confirm this, Rutherford beamed alpha rays into an evacuated double-walled glass container and found inside the chamber were helium atoms that had lost two electrons. Alpha rays were not really rays but streams of particles.

To study the penetrating power of alpha particles, Rutherford shot them at a thin piece of mica and noted that the resulting beam was much fuzzier than the original. Working with Hans Geiger (1882–1945) in Manchester in 1907, he created a detector that could record the passage of a single particle. Using this detector technique, one of his students, Ernest Marsden (1889–1970) conducted an experiment in which he shot a beam of alpha particles at a strip of gold foil. The foil was thin, only 0.008 mm thick, and most of the alpha particles passed through in a straight line. A few were deflected, and a tiny portion, about one in 8,000, bounced straight back. This was an astonishing result. Rutherford deduced that it meant that the alpha particles were striking a solid core of an atom, while the rest of the atom was essentially empty space. The solid core, which he called the "nucleus," contained almost the entire mass of the atom. Thus an atom was a nucleus with electrons around it.

That the atom was made up of separate parts fit with the discoveries of others, but that it was mostly empty space seemed to some almost impossible. The picture became even more complex when Danish physicist Niels Bohr (1885–1962) came to work with Rutherford in 1912. He demonstrated that the nuclear model could not follow classical rules of physics; otherwise the electrons in an atom would lose energy and spiral down into the core in a fraction of a second. Rather, the electrons orbited in a series of layers, and the electron's orbital momentum was "quantized," or fixed at specific values. Energy was emitted from the atom only when the electrons jumped from one orbital level to another. Further, the amount of energy (the quanta) was fixed – a single "packet" of energy was needed for a jump. (See figure 8.9.)

Here we see the study of atomic structure converging with another line of scientific research, thermodynamics. Bohr could apply this insight to the structure of the atom because Max Planck (1858–1947) had already developed quantum theory out of the study of thermodynamics. A central question in thermodynamics in the late nineteenth century was about the relationship between heat and radiation. Gustav Kirchhoff

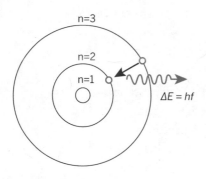

8.9 THE RUTHERFORD-BOHR ATOM

A model of the hydrogen atom. Each circle represents the possible orbital zone for an electron, called the atomic shell. If an electron moves from one orbit to another, it gains or loses a fixed amount of energy or quanta.

(1824–87) in 1859 had devised a thought experiment to visualize this relationship. He imagined a "black body" that would absorb all radiation that fell on it from infrared (heat) through visible light and on to higher energy levels such as ultraviolet. This theoretical body also worked in the other direction. If it were heated, it should radiate at all frequencies, glowing through the spectrum. Consider a brick. At 50°C it is hot to the touch, but invisible in a dark room to the human eye. At around 700°C it is just hot enough to glow, while it emits a bright light at about 6,000°C, which is about the temperature of the surface of the Sun.

A black body would be a useful experimental tool, but it could exist only in theory. In 1895 Wilhelm Wien (1864–1928) came up with a great dodge around the problem of a real black body. He reasoned that a hole in a graphite furnace could replicate a black body as closely as experimentally possible. Graphite can absorb about 97 per cent of the radiation that falls on it, which is good but not nearly good enough. By putting a hole in the graphite, radiation not absorbed would not radiate away but bounce around until it was absorbed on another surface inside the furnace. Some might escape out the hole, but most would be absorbed. Similarly, the system would work in reverse; as the furnace was heated, energy would radiate from the hole.

Wien found that, as he raised the temperature, the energy radiated at a range of frequencies but had a peak range, so that most of the radiation given off was at a particular frequency. This ran against what physicists had expected to find. Hot bodies were expected to radiate at all frequencies with equal probability, so there should have been far more high frequency radiation (such as violet and ultraviolet), because there were more high frequency possibilities than low ones. Just the opposite happened. There was a lot of low frequency radiation and very little high frequency.

In 1900 Max Planck offered a solution to the problem. He reasoned that energy was not given off continuously but in packets. He called these packets *quanta* from the Latin for "how much." Under this theory, violet light, with a high frequency, required a quantum twice as big as the quantum for red light, which has a much lower frequency. No violet light could be radiated until there was enough energy to make up the right size packet for it, but some of the energy would have already been used up in the smaller packets required for red light. Thus, red light would appear far more often than violet light, which might never be radiated at low temperatures at all. From Planck's work on the black body problem came Planck's constant h, the smallest unit of action, which is about 6.6×10^{-27} erg seconds. When he first

CONNECTIONS

Scientist and Empire: Ernest Rutherford from the Colonies

When the Right Honorable the Lord Rutherford of Nelson, OM FRS, died in 1937, he was one of the most famous and influential scientists in the world. He had won the Nobel Prize for Chemistry in 1908, been knighted in 1914, and won a number of important medals and honors in the years that followed. He served as the president of the Royal Society from 1925 to 1930, and was made a lord in 1931. What makes this doubly remarkable was that Rutherford had a colonial past. He had been born in the small town of Brightwater, New Zealand, and had done his important early work in Canada, far from the centers of intellectual power in Europe. In Rutherford's day, few people with his background were accepted into the first ranks of the British scientific elite.

Rutherford earned his BSc at the University of New Zealand. He conducted research on magnetism and developed a new form of radio receiver. In 1895 he was awarded a fellowship from the Royal Commission for the Exhibition of 1851 to do postdoctoral study at the Cavendish Laboratory at Cambridge University under the supervision of J.J. Thomson, although his doctorate would be from the University of New Zealand. He was one of the first "aliens" (those without a Cambridge degree) to be allowed to do research at the university. Rutherford was greeted with some hostility by scientists already in the lab, but he was encouraged by Thomson, who in 1900 recommended

introduced his idea about quantum theory, it did not have a huge impact, but its importance grew as other scientists applied it to a wider and wider range of physical phenomena, including Bohr's model of atomic structure.

One of the most important applications of the quantum theory was Einstein's explanation of the photoelectric effect. Hertz had discovered in 1887 that light beamed at a metal plate would eject electrons. In 1902 Philipp Lenard (1862–1947) found that there was a threshold frequency for the effect. If light was beamed at a plate below the required frequency and regardless of its intensity, no electrons would be ejected. Einstein took the idea of a packet or particle of energy (the minimum needed to break the electron bond to the metal) and in effect said that light was a kind of particle. This atom of light, or *photon*, overturned a century of theory about the wave nature of light.

him for a position at McGill University in Montreal, Canada.

In the comparatively isolated colonial world of Canadian science, Rutherford began to work on atomic structure. While in Montreal he conducted his early work on radioactivity with Frederick Soddy, creating a "Theory of Atomic Disintegration" that argued radioactive substances were transforming from a heavier substance into a lighter substance as they emitted radiation. Until that time it had been assumed atoms were indivisible. He also coined the terms for three different types of radiation: alpha, beta, and gamma rays. This work would form the basis for his Nobel Prize of 1908. An advantage of his time at McGill was that he was free to explore what he thought was interesting. The disadvantage was that resources were limited and there were few colleagues with any knowledge of advanced physics.

In 1907 Rutherford returned to England to take up the chair of physics at the University of Manchester. Manchester was an important university but did not have the status of Cambridge. At Manchester Rutherford did his famous work with Hans Geiger and Ernest Marsden, leading to the Rutherford model of the atom. He also became the first person to deliberately transmute one element into another. In 1919 he succeeded J.J. Thomson as director of the Cavendish Laboratory and became the leading physicist in Britain. Under his direction three separate teams would produce Nobel Prizes: James Chadwick for discovering the neutron, John Cockcroft and Ernest Walton for splitting the atom in a particle accelerator, and Edward Appleton for demonstrating the existence of the ionosphere.

Ernest Rutherford became one of the most influential scientists in mid-century Britain and the larger scientific world. His work led to a completely new understanding of the atom and led to splitting the atom. Without his discoveries nuclear power and the atomic bomb would not have been possible. He supervised a generation of important scientists at Cambridge and was buried in Westminster Abbey. Not bad for a boy from the South Island.

Conclusion

As the century drew to a close, the two great threads of physical science, the study of matter and the study of energy, were being knit together. The Newtonian impulse to synthesize all of nature into simple and universal laws seemed to have come triumphantly through a period of testing that arose from the problem of adding thermodynamics, electricity, and radiation to the Newtonian universe. The structure of the atom was a surprise, but for most chemists the atom's inner structure was of less importance than the clarification of atomic weights and the practical tools that were created by electrochemistry. There were a few problem areas such as the odd concept of the quanta, but the useful applications of the new knowledge far outshone any of the irksome questions still unanswered. Many scientists, in all branches of study, felt a sense of satisfaction that science had successfully answered the big questions, and some physicists actually predicted the end of physics, as all known phenomena were effectively understood.

The end of the nineteenth century was for many a time of great excitement and accomplishment. New inventions were appearing so quickly it was hard to keep up. Railways crossed the continents, steamships ruled the oceans, and telegraph wires connected the civilized world. Photography gave us images, and the phonograph recorded sounds that seemed to offer the chance to transcend old barriers of space, time, and class.

As astonishing as the new age was, there were dark clouds on the horizon for society and for science. The Crimean War (1853–6) and the American Civil War (1861–5) had given a hint of what warfare on an industrial scale could be. The rapid urbanization of the industrial countries created massive slums and rampant social problems. Although historians have argued that the urban poor were better off than the agrarian poor, especially over the long term, the "proletariat" were both more visible and more aware of their situation, in part because of the very inventions that were transforming the industrial world. Add to this the struggles over natural resources in Europe and colonial territories in Africa and the stage was set for conflict.

In science, there were a host of questions – such as how did Darwin's macrobiological system express itself on the individual level? and what did light propagate through in the vacuum of space? – that still needed good answers, but only a few scientists had any sense that the answers would not only require new insights but would transform the very foundation of scientific knowledge. To complicate matters, these changes would take place at a time when many scientists were being asked (or in some cases ordered) to turn their work to warfare and mass destruction. Although thinkers from Archimedes to Galileo had been asked by their patrons for help in times of war, the scale of the new efforts would be vast. It would also help introduce a new model of scientific research called "Big Science." The gentleman scientist working with a few dedicated assistants in a shed or university basement would be replaced by large laboratories, teams of scientists, and massive budgets. While the individual genius would still have a revered place in science, turning ideas into research would become a much bigger process.

Essay Questions

1. What led Ernest Rutherford to create a new model of the atom?

2. Explain how the kinetic theory of gases helped solve the problem of the nature of heat?

3. What experiments demonstrated the relationship between electricity and magnetism?

4. How did Röntgen discover X-rays?

CHAPTER TIMELINE

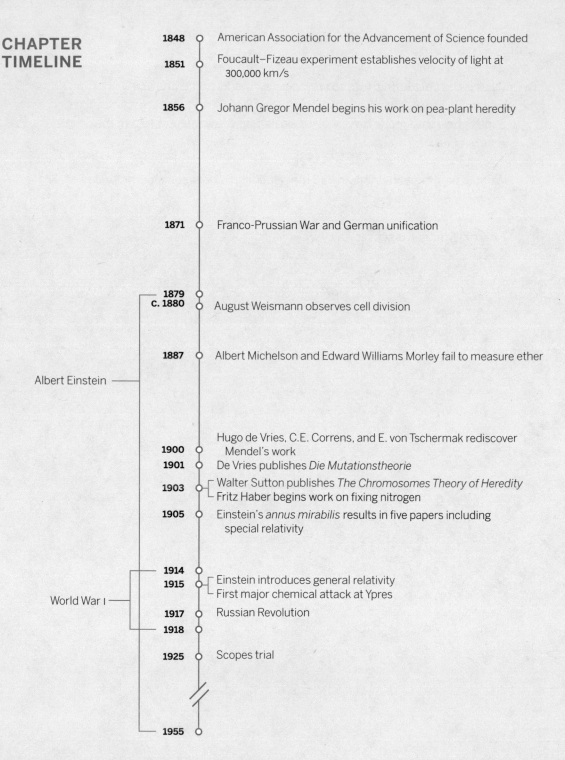

1848 American Association for the Advancement of Science founded

1851 Foucault–Fizeau experiment establishes velocity of light at 300,000 km/s

1856 Johann Gregor Mendel begins his work on pea-plant heredity

1871 Franco-Prussian War and German unification

1879
c. 1880 August Weismann observes cell division

1887 Albert Michelson and Edward Williams Morley fail to measure ether

Albert Einstein

1900 Hugo de Vries, C.E. Correns, and E. von Tschermak rediscover Mendel's work
1901 De Vries publishes *Die Mutationstheorie*
1903 Walter Sutton publishes *The Chromosomes Theory of Heredity*
Fritz Haber begins work on fixing nitrogen
1905 Einstein's *annus mirabilis* results in five papers including special relativity

1914
1915 Einstein introduces general relativity
First major chemical attack at Ypres
World War I
1917 Russian Revolution
1918
1925 Scopes trial

1955

SCIENCE
AND WAR

n many ways early nineteenth-century physicists and chemists resembled the medieval scholars who had fit their work into the Aristotelian schema even as they addressed totally new ideas. In the nineteenth century the accepted model was Newtonian. Unlike the collapse of the Aristotelian system, however, when old observations, methods, and philosophical foundations were all rejected, the new physics did not destroy all of Newtonianism or "classical physics" but rather absorbed it into a larger system. While in a strict sense Newton's laws were shown to be incomplete, they worked so well at the human level that we still live in an essentially Newtonian world. It was only in the realm of the very small and the very big that the seams of Newtonian physics and philosophy did not quite mesh. The problem for science was that the universe must be built up from the very small to the very big, so if the two ends of the scale did not follow the rules that seemed to apply in the middle, something was very wrong.

By the middle of the twentieth century science had torn apart the stable, certain, and comfortable world of Newtonian physics as it replaced axioms with probabilities. Absolute time and space gave way to observations based on the relative position and physical state of the observer, while faith in Newtonian certainty was abandoned. In biology, statistical understanding of populations replaced individual field observations as the major interpretive explanation. At the same time that science dealt these blows to Newtonianism, scientists were also aiding in the destruction of European society as, for the first time, science was directly used to support mass warfare. Science became a huge enterprise, involving substantial infrastructure and many people's labor. This new scientific and global warfare, in turn, caused many scientists and philosophers to lose their optimism and search for relativistic rather than absolute answers to human problems.

The Unfinished Business of Light

One of the most important sources of change to science came from the ongoing debate over the nature of light. Was light a wave or a particle? The supporters of the wave side were some of the few scientists to actively challenge a Newtonian idea, and their work in the nineteenth century had contributed to the rise of the Energetik position and the theory that all of nature was composed of waves and fields of force. The atomists responded by demonstrating the physical reality of atoms and even concluded that the electron, the basic unit of electricity, was a particle. To understand what was happening with rays and particles, a further tool was necessary. This was an evaluation of the velocity of light. Knowing the velocity of an object meant that a range of other properties could be calculated. From Galileo on, a number of natural philosophers had examined the velocity of light. In 1676 the Danish astronomer Ole Roemer (1644–1710) had used careful observations of Jupiter's satellites for his calculations, but his work was not widely accepted. Armand Fizeau (1819–96) and Jean Bernard Foucault (1819–68) made detailed tests of the velocity of light, finding that it traveled more slowly in water than air, and measured its velocity in a vacuum to be about 300,000 kilometers per second.

The nature of light waves and the method of propagation then became important subjects for investigation. Many scientists assumed that a "luminiferous ether" must exist through which light could travel. The term ether was borrowed from the Aristotelian name for the special matter that made up the supralunar sphere, although it was a different sort of substance than either the Aristotelian ether or the Cartesian plenum, and was similar to the concept of ether used by scientists such as Maxwell to explain electromagnetism. In fact, the problem was to identify its characteristics. The rate at which waves travel depends on the rate at which the medium can deform and return to its original state. For example, the velocity with which ripples (waves in water) move away from the point where a pebble hits the surface of a pond is limited by gravity and the molecular attraction of the water molecules. No matter how fast the pebble is traveling when it hits the water, the velocity of the waves remains constant. In other words, the speed of propagation depends on the medium. If light waves followed the same rules as sound waves, then the ether should have certain characteristics similar to those found for the propagation of sound waves in different media.

Because the velocity of light was very high, the rigidity of the ether must also be high, since the more rigid the material, the faster it can deform and snap back. That meant that the rigidity of the ether had to be higher than that of steel. There was, however, a big problem. The ether also had to be almost immaterial, or it would slow the motion of the planets, collapsing the universe in just the way Newton claimed Descartes's vortex model of the solar system would collapse. This seemed to be an attack on common sense. And yet the characteristics of the ether appeared to be the inevitable consequence of wave behavior, and the wave behavior of light had been established by the experiments of nineteenth-century physicists.

In 1881 Albert Michelson (1852–1931) created the interferometer to test the characteristics of the ether. The device used a carefully arranged set of mirrors, glass plates, and a half-silvered mirror (half the light would be reflected and half passed through) to divide a beam of light and then recombine it. The interferometer could be used to establish precisely the wavelengths of light, but Michelson had a bigger objective: he would use it to measure "ether wind."

Even though the ether was assumed to be at absolute rest, the motion of the Earth through it should create a differential rate of "flow" in the direction of the motion of the Earth around the Sun or perpendicular to it. From a mathematical point of view it was exactly the same to say that the Earth was moving through a still ether or that the ether was flowing around a still Earth. Thus, a beam of light going in one direction should not have the same velocity as a beam of light sent out at right angles, just as a boat going across a fast-flowing river does not move at the same velocity as a boat rowing up the river, even if they are being rowed at the same rate. Therefore, by constructing an experiment analogous to the boat example, Michelson hoped to measure the rate of the "current" of the ether.

In 1885 Michelson joined forces with chemist Edward Williams Morley (1838–1923) to perfect the conditions for the interferometer tests (see figure 9.1). They reasoned that if the divided beam of light traveled at different velocities because of the ether "wind," when the light was recombined it would be out of

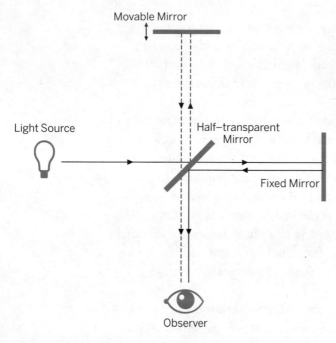

9.1 ETHER DRIFT APPARATUS FROM MICHELSON AND MORLEY'S EXPERIMENT (1887)

phase and thus would produce a noticeable pattern of fringes or areas of light and dark. They went to great lengths to eliminate all sources of error, but after many tests over many days and months they could observe no difference in the velocity of the light beams. Their experiment seemed a failure. Hendrik Lorentz (1853–1928) advanced the Lorentz-FitzGerald contraction theory to explain the failure by arguing that the equipment actually got physically shorter along the line of motion and that this contraction was just enough to put the light beams back in phase. While this was an ingenious solution, it was not a particularly satisfying one, since it suggested that it was not possible to study the ether.

Einstein's Theory of Relativity

There the matter of the ether stood until a new view of how light functioned was presented in 1905. When Albert Einstein (1879–1955) first published his work on light and relativity, he seemed an unlikely source for a series of brilliant insights in physics. His early life had been disrupted as his family moved a number of times. He did well in school, but because he was Jewish and a bit of an outsider, he did not initially aim for a research career, hoping instead to become a school teacher. After he graduated from the Eidgenössische Technische Hochschule in Zurich, he was employed in the Swiss patent office as a technical examiner; this afforded him the time to think about questions that interested him in physics since he could frequently do his day's work before lunch. Einstein was a very good mathematician (the story that he once failed mathematics in school is untrue), but his greatest strength was in conceptualizing problems in such a way that new approaches could be created to tackle them.

In 1905 Einstein had something of an *annus mirabilis* comparable to Newton's of 1666. In that year he published five papers: two on Brownian motion, two on special relativity, and one on light quantum in the photoelectric effect. His paper, "On the

Electrodynamics of Moving Bodies," looked at the motion of light as investigated by the Michelson-Morley experiment. Einstein asked himself what one would see if one were traveling on a beam of light. He made two assumptions about this motion: first, that the motion of an object must be considered relative to the motion of some other point or object. This assumption was necessary to establish the frame of reference, which was arbitrary rather than absolute as it had been in Newtonian physics.

Second, Einstein assumed that the velocity of light in a vacuum would always be the same, regardless of the motion of the light source or the motion of the observer. Picture yourself standing at the front of a speeding train. If you threw a rock forward, the rock would have the velocity of rock plus train. But if you were to shine a flashlight from the front of the train, the light would have the same velocity as light from a flashlight switched on by someone at the train station. One of the curious things about this assumption was that it did not by itself rule out objects going faster than the speed of light in a vacuum (or in more practical terms, what would happen if two beams of light were projected in opposite directions), but it explained that such objects could not be sensed by us and therefore did not really "exist" in our frame of reference.

Here Einstein extended an observation made by Galileo, who had described a situation in which a person on a moving ship dropped a rock from the top of the mast. To the people on the ship, the rock would appear to fall straight down, landing at the base of the mast. To an observer on the shore, the rock would appear to fall in a parabola. Which observer saw the "true" motion of the rock? The answer was that they both saw the true motion, but because of their frames of reference, what they saw was different, each being relative to their own motion. The concept of relativity shattered Newtonian certainty, which had been based on a single true view of action in the universe. It also undermined ether theory, which had required the ether to be stationary for all frames of reference; that is, the ether remained still while all objects (the Earth, for example) moved through it.

While the philosophical implications of Einstein's work were radical, his argument about light accorded with observation. Yet this led to some strange implications about the relationship between mass and energy, not just about the problem of frames of reference. In his second paper on relativity, "Does the Inertia of a Body Depend on Its Energy Content?" Einstein explored the effect motion would have on objects. According to the Newtonian system – $a = F/m$, where F was force, m was mass, and a was acceleration – it followed that if mass was constant (it was unchangeable by definition in the Newtonian system) and the applied force was constant, then an object could accelerate indefinitely to any velocity. However, if the velocity of light was a maximum velocity, and there was a constant force pushing the object, the acceleration of the body had to *decrease* as the velocity increased, until eventually the acceleration became zero when the object was traveling at the velocity of light. Further, if Einstein was right, then changing the Newtonian equation around to examine the mass, $m = F/a$, produced an increase in mass as acceleration decreased toward zero but the force remained constant, which violated the conservation of mass assumed in Newton's physics. Another way of saying this is that as an object traveled faster, its mass increased, but that seemed logically impossible in the Newtonian system.

The way around the problem of increasing or decreasing mass was to realize that mass was not being created or destroyed as the object moved at different speeds, but that mass and energy were different aspects of a single entity. As Joule had shown for thermodynamics, energy could not be created or destroyed, but it could be converted from one form to another. The equivalence of mass and energy had immediate application, since it helped explain the problem of radioactivity, where energy seemed to come from nowhere. What was actually happening was the conversion of mass to other forms of energy; by careful observation it was found that radioactive materials lost mass. Einstein worked out the relation of energy to mass and produced the most famous equation in history: $E = mc^2$, where E is energy, m is mass, and c is the velocity of light in a vacuum. The use of c comes from the Latin *celeritas* meaning "velocity."

If destroying the Newtonian concept of absolute motion and constant mass was not enough, Einstein's theory of relativity also destroyed absolute time. As an object went faster relative to a stationary observer, the time elapsed for the object would be different than that elapsed for the observer. In fact, the occupant of a rocket ship traveling at 260,000 kilometers per second (close to the speed of light at 300,000 kilometers per second) would experience only half the time elapsed that a stationary observer would experience. After one hour by the stationary observer's clock, the rocket's clock would show that only 30 minutes had gone by.

Einstein's 1905 work was called "special relativity" because he was examining a special case or class of relationships in which things experienced uniform motion. While a large number of objects in the universe were covered by this, special relativity did not cover all relationships. Einstein continued to work on relativity, introducing in 1915 general relativity, which included accelerated systems, such as those represented by acceleration due to gravity. Since the universe is constructed by gravitational forces, relativity covers the biggest of big pictures. In it, space-time is considered as a four-dimensional geometric construct that is uneven. The bumps, lumps, and hollows represent the gravitational distortion of the continuum.

Imagine that a sheet of rubber with a bowling ball in the center is pulled down into a cone shape. If you try to roll marbles from one side of the sheet to the other, they will curve around the cone, some of them going down to join the bowling ball rather than making it to the far side. The others will follow a curved trajectory rather than a straight line, even if they make it to the far side. Rather than saying that the bowling ball attracts the marbles, we see the marbles and the bowling ball as linked by the geometry of the rubber sheet.

General relativity linked energy, mass, time, motion, and gravity. Many of Einstein's theoretical ideas were later demonstrated experimentally, such as time dilation, in which synchronized atomic clocks went out of sync when they were moved at different velocities. Other experiments showed that gamma rays gained energy when falling into a gravity field and that the gravity of the Sun could cause light to curve. Einstein's work overturned much of the Newtonian worldview, but unlike the destruction of Aristotelian physics by Galilean and Newtonian physics, much of Newton's utility remained unchanged. In the terrestrial frame of reference, the world continued to be basically Newtonian, but Newtonianism became a special case within a larger system. What relativity did undermine was the necessity of an ether.

Relativity was not, by itself, the complete death of classical or Newtonian physics, however. Consider the form of $E = mc^2$ and $F = ma$. Although the simplicity of Einstein's equation hides a world with relative rather than absolute reference points, it also suggests a degree of certainty about the universe. There might be radically different frames of reference, clocks that moved at different times and other strange phenomena, but within a given frame of reference, you could find definite answers to physical questions. Indeed, Einstein was strongly opposed to proposals that the universe behaved in uncertain ways, famously saying, "I can't believe that God would choose to play dice with the world."[1]

Einstein's work opened the doors to the highest levels of academic research, and he held a number of posts until in 1914 he was appointed the director of the Kaiser Wilhelm Institute for Physics in Berlin and also became a professor at the Humboldt University. In 1921 Einstein won the Nobel Prize in Physics, ironically not for his work on relativity, but rather for his work on the photoelectric effect. Although Einstein was well known in the physics world, he was not yet the

....

1. Albert Einstein, *London Observer*, 15 April 1964; also in Ronald W. Clark, *Einstein: The Life and Times* (New York: World Publishing Co., 1971), 19.

international star he would become. Such considerations were overshadowed by the calamitous events of the Great War.

Mendel and the Mechanism for Evolution

While theoretical physicists explored the limits of observation, the ideas, techniques, and tools of advanced chemistry and physics had a significant impact on another area of science. Biology throughout the nineteenth century had consisted largely of field observation, classification, and anatomy. With the introduction of laboratory-based research, some biologists adopted a more experimental approach, made possible also because of new tools. These included better microscopes and cell-staining techniques (descended from Perkin's aniline dye), the use of X-rays and crystallography, and other improved methods of organic chemistry that made dealing with the sensitive materials of cells possible. The crossover of tools and techniques from chemistry and physics created modern cellular biology. When combined with statistical methods borrowed from mathematics, biologists created the "new synthesis," which produced a powerful new model of evolution by folding the microbiology and chemistry of genetics into the macrobiology of natural selection.

Biologists did not wholeheartedly support Darwinian evolution by natural selection because it lacked a mechanism for inheritance – in order for variation to be selected, that variation had to be passed to the next generation, and there was no clear way for this to happen. Darwin himself had proposed a theory called pan-genesis: each parent contributed "gemmules" to the offspring, and these gemmules somehow remained in the body to be used for reproduction when the time came. Even Darwin found this explanation unsatisfactory, but it was some time before anyone took up the challenge, largely because Darwin's theory had attracted field naturalists rather than laboratory biologists. It was not until experimentation and laboratory practice became part of the biologist's tool kit that scientists could begin to investigate inheritance at a cellular level.

Mendel's Plant-Breeding Experiments

Efforts to understand the mechanism of evolution moved more and more to the examination of the cell. The major problem was how to relate cellular events to the macrobiological result. In other words, how could biologists connect what happened inside the cell with the structure, development, and behavior of the complete organism? The first attempt to do so came before the introduction of new laboratory techniques, with a long and arduous plant-breeding experiment undertaken by an Augustinian monk in Moravia. Johann Gregor Mendel (1822–84) was the son of Silesian peasants. He got his education as many poor but bright boys did by going to a monastery school. He later joined the Augustinian Order at Brünn and went on to study at Vienna University. He began his painstaking work on plant heredity around 1856. By the time he finished his research, setting it aside when he became abbot, he had bred and examined more than 28,000 pea plants. He identified seven characteristics that had two distinct forms and that bred true from generation to generation. (See figure 9.2.)

9.2 MENDEL'S SEVEN CHARACTERISTICS

1.	Form of ripe seed
2.	Colour of seed albumen
3.	Colour of seed coat
4.	Form of ripe pods
5.	Colour of unripe pods
6.	Position of flowers
7.	Length of stem

He then crossbred plants with alternate forms and found that the traits did not blend but remained discrete. Further, he discovered that successive generations of plants followed a clear pattern in the appearance of characteristics in the offspring. Some characteristics were dominant and others recessive. In other words, if a plant with round seeds (dominant) was crossed or bred with a plant with bumpy seeds (recessive), all the offspring in the first generation would have round seeds. But the recessive characteristic did not disappear; it was just not expressed in the shape of the plant. In the next generation, breeding two hybrid plants produced a mixture of round and bumpy seeds, as a certain percentage of offspring received dominant characteristics and a smaller number received only recessive characteristics. The ratio of dominant to recessive appearance worked out to 3:1, but in terms of the distribution of the two characteristics it represented four genetic combinations: round/round, round/bumpy, bumpy/round, and bumpy/bumpy. By crossing or hybridizing plants with different characteristics, Mendel demonstrated an algebra of inheritance to explain the transferral of characteristics. For two characteristics, the distribution was 9:3:3:1. (See figure 9.3.)

Mendel's work identified two key concepts which have sometimes been called the law of segregation and the law of independent assortment. The law of segregation identified the link between inherited characteristics and sexual reproduction. Two parts, one from each parent, were needed to create a characteristic. Those two parts had to be the same type of material but could be different versions. In

First hybrid crossing a round seed with a bumpy seed

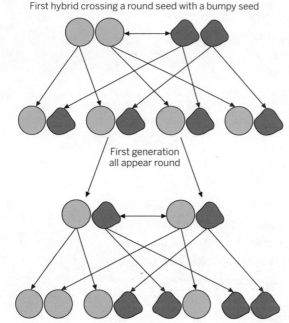

First generation
all appear round

Second generation of hybrids produces a 3:1 ratio of round to bumpy

9.3 MENDEL'S SEED ALGEBRA

reproduction each parent contributed exactly half of the material, which meant that whatever it was in the cell that controlled the characteristic had to segregate during gamete (sperm or egg) production. Because of the distribution of characteristic-producing material, the law of independent assortment said that variant characteristics remained independent of each other (as the algebraic illustration makes clear), and thus there were fixed ratios of characteristics in the offspring. Mendel's work explained atavism (throwbacks) and disproved the idea of blended inheritance. It had nothing to say about evolution and, indeed, seemed to make evolution less possible, since discrete units of inheritance did not seem to change.

Mendel read his paper describing this research at the Brünn Natural Scientific Society meeting in 1865, and it was published in the society's journal, but the paper had almost no impact on the biological research community. The copy of the paper that Mendel personally sent to Darwin went unread, remaining unopened in Darwin's library. The reasons for this lack of recognition are not hard to understand. Mendel was a monk in a monastery far from the main centers of scientific activity and so was not known to the biological community. The hot topic in biology was evolution, but his work looked like plant breeding, a branch of horticulture. Moreover, the results Mendel found did not seem transferable to other species. Mendel sent his results to the foremost German botanist, Karl Wilhelm von Nägeli (1817–91), who questioned his findings and suggested he work on the plant hawkweed. Because hawkweed does not breed true, it failed to show the pattern that the pea plants had exhibited and so challenged the universality of Mendel's results. Further, there was no material basis for Mendel's statistical theory. What were these units of inheritance?

Mendel's work offered a method for predicting how characteristics were distributed in a population over multiple generations. It also directed attention to the equal importance of material communicated from each parent to the offspring. While it had been recognized for generations that reproduction was the key to understanding the likely characteristics of offspring, there was much confusion about the proportion of contribution in sexual reproduction, ranging from the female as mere vessel for the creative material of the male-to-male characteristics passing only to males and female characteristics only to females. Mendel's work demonstrated that in sexual reproduction each parent contributed half of the matter and information necessary to create a new organism.

During the 1880s August Weismann (1834–1915) attempted to disprove Darwin's pan-genesis theory through a laboratory demonstration of reproduction, using microscopic observation of cell division. Weismann argued that cells were immortal (since one-celled organisms reproduced indefinitely through division) and that the cell nucleus had the main role in passing information about genetic traits from parent to offspring. He observed the longitudinal split of the nucleus in mitosis (the process of division of a cell into identical pairs), which suggested that the germ plasm (as he called the material substance of inheritance) resided in the nucleus. Since sexual reproduction resulted in the combination of germ plasm from each parent, variation was produced through the many possible chromosomal combinations. The logical conclusion from these findings was that heredity was based exclusively on internal biological factors (hard heredity) rather than on any environmental influence or the inheritance of acquired characteristics.

Despite, or perhaps because of, these investigations, the link between the behavior of cells at the microscopic level and the evidence of evolution at a macroscopic one was still unclear until Mendel's work was rediscovered in 1900. Three researchers, Hugo de Vries (1848–1935), C.E. Correns (1864–1935), and E. von Tschermak (1871–1962), all working on rates of variation in plant populations, discovered Mendel's work and recognized its importance. De Vries labeled the unit characteristics of heredity "pangens," which was soon shortened to "genes." He also recognized that although Mendel's system allowed variation through generations, it would require major shifts if evolution was to occur. He called the sudden changes "mutations" in his 1901 book *Die Mutationstheorie* and suggested that this might be a way to understand the introduction of new characteristics. In 1903 Walter Sutton's (1877–1916) book *The Chromosomes Theory of Heredity* narrowed the target area of genetic activity. He argued that genes are carried by chromosomes and that each egg or sperm cell contains only half of the chromosome pair. In effect, de Vries and Sutton discovered the cellular system predicted by Mendel's work.

The chromosome was not exactly discovered as a single moment of insight, but rather was observed by a number of people doing cell biology in the nineteenth century. Matthias Jakob Schleiden (1804–81), Rudolf Virchow (1821–1902), and Otto Bütschli (1848–1920) all noted this intracellular structure that was made observable by better microscopes and the introduction of cell stains from artificial dyes. Walther Flemming (1843–1905) observed that the nucleus of a cell split during cell division and theorized that the nucleus was passed on from generation to generation, saying "*omnis nucleus e nucleo*" ("every nucleus from a nucleus"). Flemming did not observe the chromosome split equally, missing a key point in its role in heredity. The chromosome was named by Heinrich Wilhelm von Waldeyer-Hartz (1836–1912), combining the Greek word for color (chroma) and body (soma), since the chromosome strongly absorbed dye.

Two scientists, Theodor Boveri (1862–1915) in Germany and Walter Sutton in the United States, independently demonstrated the chromosome theory of inheritance. Following the publication of Edmund Beecher Wilson's (1856–1939) *The Cell in Development and Heredity* (1902) that promoted the Boveri-Sutton chromosome theory, the chromosome became the target of intense study. In 1923 Theophilus Painter (1889–1969) published his observation that humans had 48 chromosomes. In 1956 the number of human chromosomes was reduced to 46 when better tools allowed Joe Hin Tjio (1919–2001) and Albert Levan (1905–98) to do a more detailed examination. This in turn led to the discovery of genes, which were specific sections of chromosomes. This was first demonstrated by Thomas Hunt Morgan (1866–1945) in 1910 and led to chromosomal "maps" that linked the physical characteristics of an organism to specific locations on the chromosomes.

Although the chromosome was established as the cellular body that transmitted the information of heredity from generation to generation, it would take another 40 years to decode the chemical basis of how a gene worked.

Science and War

When the Great Powers met at the Berlin West Africa Conference in 1885, the results made war almost inevitable. In many ways the conference marked the apex of colonialism, as the major Western nations met to discuss trade and navigation and to divide up the unclaimed portions of Africa. (See figure 9.4.) For Germany it was a test of its rising status as an economic and political superpower. Although it had only become a unified state in 1871, Germany was already challenging the industrial might of Britain and sought to establish colonial holdings in accord with its status as a European power. Germany had been late entering the race for

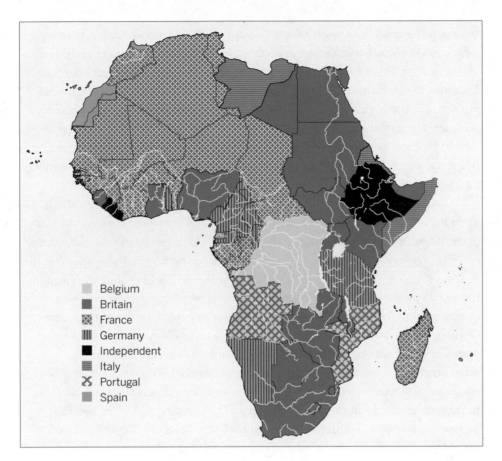

9.4 AFRICA AFTER THE BERLIN CONFERENCE, 1885

Legend:
- Belgium
- Britain
- France
- Germany
- Independent
- Italy
- Portugal
- Spain

colonies, so much was at stake. International status, access to resources, military advantage, and political benefits could be derived from colonial holdings. The conference did not produce the gains Chancellor Otto von Bismarck had hoped for, however. Although Germany did annex Tanganyika and Zanzibar, these territories were far less valuable than those held by small states, such as Belgium's control of the Congo or Portugal's colony in Angola.

Germany faced a difficult situation as a European power. It was bordered on the east by the "sleeping bear" of Russia with her huge territory, limitless natural resources, and growing industrial base. On the west was France, which was less industrialized than Germany but had better natural resources, access to two oceans, and significant colonial territory. There was much antipathy between the two countries not least because of the Franco-Prussian War, in which France was badly beaten and which led to the unification of the German states in 1871. France was forced to give up resource-rich Alsace and parts of Lorraine to Germany. Dominating the seas and holding the greatest colonial empire was the highly industrialized Britain. Despite long-standing tension with France, Britain became Germany's main opponent, and it was in opposition to British sea power that Germany embarked on a massive naval arms build-up, which resulted in the first modern arms race, as each side built more and bigger ships.

To successfully pursue its imperial policy, all aspects of German society had to be harnessed together, from the farmer in the field to the captain on the bridge of the dreadnought. The interest of the state and the interests of business were often synonymous, and when Bismarck introduced the first welfare programs in Europe, it was not because of socialist leanings but rather to protect the German industrial workforce and keep it as productive as possible.

To redress the lack of resources or high cost of exploiting domestic sources, Germany turned more and more to science and technology. It created an integrated system that brought together schools and universities, industry, government, and research scientists. The new approach to science had the educational system as its foundation. The government constantly promoted the improvement of education, and by the turn of the century Germany had the highest literacy rate in Europe. Technical high schools and colleges produced skilled workers for the industrial and business sectors, while advanced education was encouraged in the many universities, leading to a huge growth in the number of professionally trained scientists.

The scientific structure linked three major entities: government (including the military), business, and academic research institutions. The government funded education and created a series of elite research centers, particularly the Kaiser Wilhelm Institutes. The education system was designed to stream students, with the best science students going on to university. The best of those students went on to graduate work, and to facilitate this the German universities introduced the modern PhD (Doctor of Philosophy), which was first awarded at Humboldt University shortly after its founding in 1810. The PhD required not only advanced knowledge of a subject but demonstrated ability to do original research. Of this elite group, the best were recruited to the research centers, while the next tier became university professors or were recruited by industry.

The universities and research centers in turn worked on problems of importance to the country. Although scientists, especially those at the top, were not commanded to work on particular problems, there were both formal and informal discussions about what topics would help the nation and what scientific work might have industrial applications. Thus, national interests, business interests, and scientific research were combined. For example, the electrical research of people such as Hermann von Helmholtz (1821–94) and Hertz was on the cutting edge of theoretical physics, but the principles were quickly converted into material suitable for industrial use by engineers. In turn, experienced engineers and scientifically trained technicians were snapped up by the growing industries of Germany.

These formal and informal ties were also maintained by a number of specialized scientific societies. The BAAS and the American Association for the Advancement of Science (founded in 1848) had tried to be umbrella organizations for like-minded men of science, but it was on groups such as the German Physical Society (1845) and the American Chemical Society (1876) that scientists increasingly relied for professional and community activities. Unlike the older Royal Society in Britain or the Académie des Sciences in France, the new societies were composed of working scientists within a single discipline. They not only provided a conduit for scientific research through conferences and journals but also became a vital resource for professional development on an individual level, by helping people get jobs, and on a community level as they advised governments, lobbied for standards and laws, and promoted the discipline.

The Chemists' War

Fritz Haber (1868–1934) and Carl Bosch (1874–1940) were two scientists whose experience typifies the intersection of national interest, industry, and research. Because of the rise in German population, pressure on agriculture was increasing, but intensive agriculture exacerbated the problem of soil depletion. Germany did not have secure access to natural sources of nitrates for fertilizers, which came primarily from guano, or bird droppings. The chief source at the turn of the century was Chile, which in 1913 produced about 56 per cent of the world supply of nitrates. To make matters worse for Germany, British companies controlled most of Chile's nitrate production. In 1903 Haber began to investigate methods to "fix" atmospheric nitrogen by converting it to ammonia, which could then be used for fertilizer as well as providing a useful feedstock for other nitrogen products. The basic process created the reaction $N_2 + 3H_2 = 2NH_3$, but the contemporary method of production required large amounts of electricity, which made it uneconomical. By 1909 Haber and Robert Le Rossignol had worked out a continuous flow method to produce ammonia; it required a reaction vessel that could withstand 200 atmospheres of pressure and be heated to 500°C. Haber convinced the chemical company BASF that the process could be made commercially viable, but to put the system into place BASF had to turn to the steelmaker Krupp to manufacture the reaction vessels. Krupp, in turn, went to its scientists and engineers to pioneer a new method of forging steel to create a container that could withstand the pressure and heat.

When all the elements were put together, the Haber-Bosch system produced large quantities of inexpensive ammonia. The synthesis of ammonia earned Haber the Nobel Prize in Chemistry in 1918 and led to the creation of the Oppau and Leuna Ammonia Works, the first industrial producer of nitrate products that did not require large-scale electrical supplies to operate. By 1934 almost 64 per cent of fixed nitrogen was produced synthetically, and the leading producer was Germany, while Chile's share of world production fell to just over 7 per cent.

Although the initial concern had been to create a secure domestic source of artificial fertilizer and to supply industrial demand for ammonia, with the outbreak of war, when natural sources were cut off by Britain and her allies, the Ammonia Works provided the nitrates necessary for explosives such as cellulose nitrate. Without the synthesis of ammonia, Germany could not have continued to fight the war for more than a few months.

Germany entered World War I with expectations that it would defeat France and her allies quickly in a repeat of the Franco-Prussian War. This did not happen. French resistance was greater than expected, the British bottled up the German navy and blockaded her ports, and the war with Russia divided German forces. When both sides tried to outflank the other, the front lines stretched longer and longer until the western front ran from the Alps to the English Channel and the eastern front stretched from the Black Sea to the Baltic Ocean. Germany had created a well-trained and literate military backed by the second most powerful industrial economy in Europe, but it could not withstand a war of attrition, especially when the British and French continued to import supplies from their colonies and the United States.

9.5 THE WESTERN FRONT, WORLD WAR I

Fritz Haber persuaded the German High Command to try a chemical attack. Haber arranged to transport 168 tons of chlorine gas in cylinders to the front and then waited for the wind to blow in the right direction. On April 22, 1915, the German army released the gas along a 6.5 kilometer section of the front line at Langemarck, near Ypres in Belgium. The yellowish-green cloud drifted across no man's land and onto the Allied position. The front line collapsed, and German forces followed the cloud through the hole in the line. Resistance behind the front lines and a lack of German reserves to carry the attack farther halted their advance. By the time the gas cleared, 5,000 soldiers were dead and 15,000 had been wounded in the attack. Haber was made the director of gas warfare after this demonstration of its effectiveness.

CONNECTIONS

Chemical War: Science and the State

World War I changed both the ideology of science and the way science was organized. Science went from being seen as one of the intellectual foundations of the Enlightenment project for reform to a morally neutral and disinterested enterprise. Although science could be used for good or evil, science itself had no moral value. The war also led to the creation of Big Science, where scientists worked in teams, operated in large laboratories, and were funded by the state or corporations at levels unheard of before the war.

Prior to 1915 most scientists believed that science was above political disputes and ideologies. Many scientists accepted the Enlightenment ideal of science as a path to truth and rationality, and thus as an active tool to create a better world for everyone. Individual scientists might be partisan, but science almost by definition could not be. It would be like saying that the force of gravity was English or the theory of acid formation belonged to France. Scientists believed participation in science was open to all and knowledge could not be constrained by legislation; anything one scientist discovered could be found by other scientists because nature was universally available for study.

This view of science began to crumble on April 22, 1915, at Ypres, Belgium, when German troops released 168 tons of chlorine gas that

Chlorine was a good choice for a weapon. It was available in large quantities because of its use in chemical and textile industries, and it was slightly heavier than air so it settled in hollows and trenches. It did its damage by combining with moisture in the eyes, nose, and lungs to form hydrochloric acid. Although it was not the first use of chemicals on the battlefield (earlier in the war, the French had used tear gas, and the Germans had ineffectively used chemical-filled artillery shells against the Russians), Ypres was the first successful gas attack, and it changed the conduct of the war. It also helped change the course of science.

swept across the Allied front line, killing at least 5,000 men and injuring more than 15,000. Haber believed that it was his duty to work for Germany during a time of war. He hoped that his chemical attack would break the stalemate of trench warfare and lead to the end of the war. His actions were not driven by a desire to kill and wound thousands of people, but to end the war as quickly as possible and stop the killing. This rationale would be applied to nearly every new weapon of war; it had been applied to the machine gun and dreadnoughts, and it would be deployed again as justification for aerial bombing and the atomic bomb.

The use of chemical warfare was the first time that science played a major role in warfare. The gas attack set off an arms race, but it also set off a brain race as each combatant nation struggled to organize its scientists for war. Before the war, only France and Germany had significant state funding of science. Because of the war Britain, the United States, Canada, Australia, and Italy established national research councils and began to fund scientific research. Scientists became a strategic resource and science was turned to the pursuit of mass destruction. The structures for scientific research that exist today were born out of that theater of war.

The ideology of science was also changed. No longer was science associated with Enlightenment ideals of rationality, democracy, and human rights. Instead, scientists began to argue that science was morally neutral – it was neither good nor bad. Science could be turned to good or bad uses, but had no inherent morality. Haber expressed this idea by saying that a scientist should serve mankind in times of peace but the fatherland in war. Scientists around the world made a kind of Faustian bargain: in return for support for scientific research at levels far beyond those of the nineteenth century, most scientists gave up the idea that science was morally uplifting. Further, they tacitly agreed that their services could be called upon for political or military needs, rather than to uplift the whole people.

All sides in the conflict flung themselves into a concerted effort to create new chemicals to use on the battlefield and defensive equipment to ward off enemy attack. What had been a contest of heavy artillery, machine guns, and trench warfare now became a battle of scientific prowess. Britain, France, Canada, the United States, and Italy created national research councils. These organizations were intended to coordinate research efforts with industrial production and military needs. In short, the Allies attempted to emulate the integration of research that Germany had already pioneered. Hundreds of chemists, chemical engineers, biologists, physicians, engineers, physiologists, and others were pressed into service. Crash programs set up laboratories, and scientists began their work even before construction crews were finished. The Allies launched their own gas attack on September 25, 1915, at Loos. It was not a great success. While part of the German front line retreated, some of the gas blew back on the British and French line. The territory gained was retaken in a few days.

Chlorine was relatively easy to recognize and defend against, so Haber and his fellow chemists had to introduce more potent chemicals if they were to keep their advantage. They turned to phosgene, which was clear, colorless, and odorless. It acted more slowly than chlorine but was more subtle, injuring or killing hours or even days after exposure, and it added a new level of terror to gas warfare. It was followed by other chemicals including chloropicrin, mustard gas, and diphenylchlorarsine. By the spring of 1916 both sides were using chemicals as standard weapons. Against an untrained or unequipped enemy, gas was deadly. It was used to great effect by the Germans against the Russians and the Italians. Against a trained and prepared force, however, it did not change the stalemate in the trenches, although it did have a profound psychological effect on both soldiers and the civilian population. The advent of chemical weapons and the development of long-range aircraft made potential targets of cities and vulnerable civilian populations far from the front. It also brought home to the governments and general public the power and importance of science and scientists.

In addition to chemical warfare, scientists were pressed into service working on almost every aspect of war from new forms of explosives, aerial combat, sonar, radio communication, meteorology, and medical treatment, including the use of the recently discovered X-ray.

The signing of the Armistice on November 11, 1918, which ended World War I, marked a turning point in European and world history. A generation had been devastated by the war and by the influenza pandemic that followed. A generation of young men had been slaughtered in a largely futile struggle. More than 8 million people were killed and over 20 million wounded. To add to the calamity, somewhere between 50 and 100 million people died in the worldwide influenza pandemic that raged from 1918 to 1920, with the European nations losing around 3 per cent of their populations to the disease. European economies lay in ruins, and Russia was still in the throes of the revolution that had started in 1917. The old political systems, the old ways of living, had changed forever. France, humili-ated by its defeat in the earlier Franco-Prussian War and reeling from the destruction of a four-year conflict fought largely on its soil, sought to punish the Germans. The terms of the 1919 Treaty of Versailles forced the Germans to give up colonial holdings and make huge reparation payments, imposed trade restrictions, and dismantled parts of its government and military. German society was thrown into turmoil, exacerbated by economic depression and hyperinflation. In 1923, when Germany was unable to make reparation payments, the French invaded the Ruhr district, taking over coalfields and industrial facilities in an armed occupation.

Science in the National Interest

The war had brought the utility of science to the forefront, as well as demonstrating its dark and dangerous side. Generations of philosophers and scientists had argued that science would benefit society and help nations prosper; now, its power and utility had been graphically illustrated by the "Chemists' War." Even as the victors tried to settle back into a pre-war life, the industrialized world could not ignore the system that had allowed one nation to challenge and almost defeat a coalition of larger and better supplied states. In the years following World War I, scientists worked to establish their national status and rebuild their network of international contacts. Almost in defiance of the political divisions appearing everywhere, scientists published in internationally read journals and worked both in competition and cooperation with laboratories all over Europe and the United States. The experience of big labs and big projects, which had been so important to war work, led scientists to look for projects that might fit that model. As physics emerged as the pre-eminent scientific discipline of the 1920s and 1930s, scientists in other areas turned to using methods, theories, and models from physics as they investigated evolution and inheritance at a molecular level, looking at cellular structure and DNA. At the same time, physicists themselves were delving ever further into the subatomic realm, with ultimately explosive results for the atomic age. Before the war, in Britain and the United States in particular, science as a profession had had a lower status than other academic pursuits, and with the exception of a few research centers such as the Cavendish Laboratory or Cornell University, most students seeking advanced training had traveled to Germany. When the war closed off this avenue of study, domestic programs had to fill the gap. And, as the war progressed, it became apparent to governments and educators that science education was a valuable national resource. The change helped the Unites States rise to scientific power.

In Germany, national interest, combined with the sense that science and technology could help the German situation, made research an exception to the generally dismal circumstances of the postwar years. Perhaps because money was in short supply for laboratories and equipment, a host of powerful minds in the universities and research centers flourished, doing advanced theoretical work. Max Planck, Werner Heisenberg, Erwin Schrödinger, Albert Einstein, Leo Szilard, Otto Hahn, and Lise Meitner were only a few of the people who did pioneering work in Germany after the war. International contacts were quickly re-established. Scientists on both sides of the conflict had felt a duty to work for their countries during the war, but they also felt a duty to the community of science. Many went further, arguing that science and the rationality it embodied would bring peace to the world, in an echo of the Royal Society's search for a "third way" in the seventeenth century.

Conclusion

Science, which had contributed to the marvels of the Victorian era, bringing new devices and, for many, greater comfort than any previous age, now showed a dark and dangerous side. It was not a coincidence that the certainty of the Newtonian worldview crumbled at the same juncture as European society was rent by a terrible war. The infrastructure that made the esoteric research of Rutherford, the Curies, Max Planck, and Einstein possible also made chemical warfare, high explosives, and aerial bombardment the face of modern warfare.

Although governments cut funding to science after World War I, the success of scientific research for the war effort made it impossible to return to the polite Victorian era of the gentleman scientist. In the years between the two world wars scientists themselves worked diligently to promote large-scale research programs and were ready to heed the call to arms when it came again. Science had been brought to public attention. The social status of science improved, as indicated by the increase in science education, particularly at the advanced level. Governments began to pay attention to the power of scientific research, often seeking to bend it to commercial, industrial, or military ends in the years to come.

Essay Questions

1.   What was the Michelson-Morley experiment supposed to demonstrate?

2.   How did Einstein's theory of relativity challenge the Newtonian system?

3.   Explain how Mendel's seed algebra supported the theory of evolution.

4.   Why was World War I also known as the "Chemists' War"?

CHAPTER TIMELINE

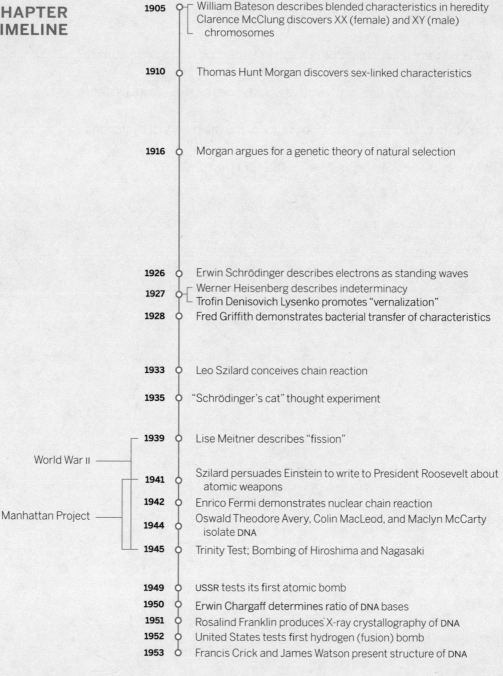

1905 William Bateson describes blended characteristics in heredity
Clarence McClung discovers XX (female) and XY (male) chromosomes

1910 Thomas Hunt Morgan discovers sex-linked characteristics

1916 Morgan argues for a genetic theory of natural selection

1926 Erwin Schrödinger describes electrons as standing waves
1927 Werner Heisenberg describes indeterminacy
Trofin Denisovich Lysenko promotes "vernalization"
1928 Fred Griffith demonstrates bacterial transfer of characteristics

1933 Leo Szilard conceives chain reaction

1935 "Schrödinger's cat" thought experiment

1939 Lise Meitner describes "fission"

World War II

1941 Szilard persuades Einstein to write to President Roosevelt about atomic weapons

1942 Enrico Fermi demonstrates nuclear chain reaction

Manhattan Project

1944 Oswald Theodore Avery, Colin MacLeod, and Maclyn McCarty isolate DNA

1945 Trinity Test; Bombing of Hiroshima and Nagasaki

1949 USSR tests its first atomic bomb
1950 Erwin Chargaff determines ratio of DNA bases
1951 Rosalind Franklin produces X-ray crystallography of DNA
1952 United States tests first hydrogen (fusion) bomb
1953 Francis Crick and James Watson present structure of DNA

THE DEATH OF CERTAINTY

10

I n the wake of World War I, long-standing social, political, and economic standards were challenged and overturned. In politics, the power of monarchies declined and new ideologies such as modern liberalism, industrial capitalism, democratic socialism, and communism gained adherents. Many democratic countries started extending voting rights to women: Canada in 1917, Britain and Germany in 1918, the Netherlands in 1919, and the United States in 1920, although French and Italian women would have to wait until 1944 and 1946, respectively, for the right to vote. The Russian Revolution changed the diplomatic map of the world.

In the arts the Impressionist movement, led by people such as painters Claude Monet and Edgar Degas, that had so shocked the cultural establishment in the 1870s and 1880s gave way to the even more radical art of Expressionists such as Wassily Kandinsky and Franz Marc and the "anti-art" art of the Dadaists such as Marcel Duchamp and Hannah Höch. In architecture the modernist movement was using cutting-edge technology and a rejection of ornamentation to create buildings for an industrial age. Charles-Édouard Jeanneret, better known as Le Corbusier, said, "A house is a machine for living in."[1] As well as the visual arts, dance, theater, film, and literature were all marked by a period of experimentation in the first part of the twentieth century. If these disparate arts had any common ground, it was two things: an attack on the old forms and a desire to look beneath the surface of things.

....

1. Le Corbusier, *Vers une architecture* (Paris: G. Crés et C., 1923).

The postwar years were a period of decadence for some people, as they tried to forget the horrors of the war, and of depredation, especially after the collapse of international finance in 1929 and the start of the Great Depression. It seems more than a coincidence that just when the stability of government and finance was shaken by rapid technological change, war, and depression, so many artists were rejecting old ideas. It is sometimes hard to see the outside world in the world of science, where Einstein's theory of general relativity appeared in 1915 during the course of the war and Pluto was discovered in 1930 during the depths of the Great Depression, but turmoil in science was just as much a challenge to the old world-view. Two areas in particular would force a reconsideration of how nature worked. The first was the continued development of nuclear physics, in particular the problem of indeterminacy, and the second was the discovery of genetics and the synthesis of evolution and biochemistry. Nuclear physics would open the door to nuclear power and nuclear weapons, while evolution and genetics would challenge social and religious ideas.

The New Physics: Indeterminacy

Although World War I was dubbed the "Chemists' War," chemistry was giving way to physics as the premier area of study. Chemistry and physics were not separate subjects at the beginning of the nineteenth century, but by the start of the twentieth century the distinction had become increasingly institutionalized and the territory more clearly defined. Chemistry focused more and more on the atomic and molecular level, while physics looked at the extreme ends of observation – the subatomic realm at one end and the structure of the universe at the other. These investigations undermined what came to be called "classical" physics, which was characterized less by its subject matter or even its experimental methodology than by its assumption of the absolute condition of nature and the resulting certainty of its laws. Prior to relativity and quantum theory, the laws of nature were considered to be simple, universal, and invariant. The location of the event or the position of the observer made no difference. The new physics removed the comfortable certainty of the old and replaced it with a more precise but contingent system. The condition of the observer became crucial to the answers one got from observation of the physical world.

At its most basic level, indeterminacy is easy to understand. Take, for example, what seems like the important but mundane task of creating a star catalog. The catalog identifies stars by a set of coordinates that allows astronomers to look at the same object in the sky. Such catalogs have existed since at least the time of Ptolemy. But there is a problem. Every telescope is different, and even with the same telescope astronomers know that things such as temperature and atmospheric conditions (for example, air density and amount of water vapor) make each observation slightly different. So astronomers make multiple observations of a star and combine the results to statistically create its coordinates. Although any astronomer using the catalog will certainly be able to find the desired object, is the star actually at those coordinates? The answer is that we don't know, and the problem is that we can't know. No physical instrument can be perfect, and no instrument can observe the universe from outside the universe and thus must interact in some way with the thing being observed. For astronomers – and essentially for anything above the atomic level – this indeterminacy has little practical effect. At the subatomic level, it created a serious scientific problem.

As in the case of the death of the phlogiston theory, rather than established physicists converting to the new view it took a new generation of scientists to develop and embrace the new ideas of quantum physics. This gulf was widened by the introduction of wave mechanics within quantum physics. Bohr's quantum picture of the atom, which seemed to work well for individual atoms, broke down in more complex arrangements, such as the simplest form of a diatomic molecule (for example, atmospheric oxygen O_2). Louis de Broglie (1892–1987) in France, Erwin Schrödinger (1887–1961) and Werner Heisenberg (1901–76) in Germany, and Paul A.M. Dirac (1902–84) in Britain all tackled the problem. De Broglie argued that waves and particles were two aspects of a single entity. Thus, electrons and photons could have particle properties if looked at one way and wave properties if looked at another. While the wave-particle duality flew in the face of about 300 years of argument about particles and waves of light, it worked neatly, particularly since anything that moves can be mathematically described as a wave.

In 1926 Schrödinger argued that electrons could be described as standing waves with modes of vibration analogous to the vibrations of violin strings. While this allowed a better mathematical description of the behavior of the electron, it was not universally accepted, as both Heisenberg and Dirac offered alternative models that were in a sense "purer" since they were strictly mathematical. What linked the wave models was the necessity of dealing with electrons and other particles as a series of probabilities rather than as completely knowable objects. This had profound philosophical consequences, not the least of which was the rejection of aspects of quantum theory by Einstein and Schrödinger, who each felt that there must be some ultimate reality in the universe, not simply a range of possible realities. They were both concerned that limits to observation constrained what could be absolutely known about the very structure and function of the universe.

Heisenberg formalized this problem in 1927 when he reasoned that it would not be possible simultaneously to know the exact position and the momentum of an electron as it orbited the nucleus of an atom. Any method that could be used to determine the position of the electron would have to interact with the electron, and that would alter its momentum. Similarly, anything that could record the momentum would interfere with the motion of the electron and change its position. This was the principle of indeterminacy, and Heisenberg expressed it as a mathematical formula: $\Delta p \times \Delta x \sim h$, where p was the uncertainty of the momentum, x was the uncertainty of the position, and h was Planck's constant. Planck's constant was a measurement of quantum energy, or the smallest "packet" of energy. Another way to think of this was to say that if Δp for an electron was completely known, then Δx would be completely unknowable, but the whole system would have a value equal to the smallest energy packet for the system.

The principle of indeterminacy has also been called the "uncertainty principle," although this is not an accurate label. While it might not be possible to determine the position and momentum of a single electron, the behavior of the electron is consistent. In other words, the electron's behavior is not uncertain in the sense of being random, but we cannot determine certain things about it because we are part of the same physical universe. Further, since we can be as completely certain about the behavior of electrons as a class of objects as it is possible to be certain about anything that can be examined in the physical world, the indeterminacy principle does not make physics (or science in a wider sense) impossible or unreliable. In fact, just the opposite. The probabilistic approach to physics is more precise than classical physics since it describes a system that can be in more than one state. In other words, it models the real world better than the ideal world assumed by Newtonian or classical physics.

Indeterminacy was disturbing because it seemed to open physics to a kind of strange metaphysics. Indeterminacy can be read in such a way that all possible conditions exist simultaneously. In 1935 Schrödinger tried to illustrate the difficulty of taking indeterminacy to its logical conclusion in an essay describing conceptual problems in quantum mechanics. In this article he offered a thought experiment that has become known as "Schrödinger's cat":

One can even set up quite ridiculous cases. A cat is penned up in a steel chamber, along with the following diabolical device (which must be secured against direct interference by the cat): in a Geiger counter there is a tiny bit of radioactive substance, so small that perhaps in the course of one hour one of the atoms decays, but also, with equal probability, perhaps none; if it happens, the counter tube discharges and through a relay releases a hammer which shatters a small flask of hydrocyanic acid. If one has left this entire system to itself for an hour, one would say that the cat still lives if meanwhile no atom has decayed. The first atomic decay would have poisoned it. The Psi function for the entire system would express this by having in it the living and the dead cat (pardon the expression) mixed or smeared out in equal parts.[2]

In other words, because we cannot know the state of the cat until we look, the cat is equally alive and dead *at the same time* since there is a 50/50 chance of either condition. Only by opening the diabolical device will the universe resolve into one state or the other.

The simple idea that the act of observation changes the thing observed found general acceptance since there could be no tool of observation that did not have physical properties such as mass or use waves or particles in its operation and thus interacted with the thing observed. Philosophically, the concept of the participation of the observer in the thing observed, the end of complete objectivity, had far-reaching implications, helping to produce the cultural relativism that transformed the social sciences in the twentieth century. Literature, anthropology, and history all began to take into account the reader, the anthropologist, and the archivist as part of the system rather than outside and unbiased observers. The anthropologist Franz Boas (1858–1942), who earned a doctorate in physics before turning to the study of human culture, rejected the idea that there was a hierarchy of civilizations and strongly opposed the scientific racism of people like Galton. Boas argued that there were no higher or lower races or cultures and that all people see the world through the lens of their own culture. There was no "objective" measure of culture, only contingent observations. Activities that are considered moral in one society might be considered immoral in a different society,

2. E. Schrödinger, "Die gegenwartige Situation in der Quantenmechanik," *Die Naturwissenscaften* 23 (1935): 807–49; John D. Trimmer, trans., "The Present Situation in Quantum Mechanics: A Translation of Schrödinger's 'Cat Paradox' Paper," *Proceedings of the American Philosophical Society* 124, no. 5 (1980): 323–38. Also in *Quantum Theory and Measurement*, eds. J.A. Wheeler and W.H. Zurek (Princeton, NJ: Princeton University Press, 1983), 152–67.

and had to be understood in context of the culture. This came to be known as "cultural relativism."

Still, for scientists to conclude that things do not have a specific state of existence unless they are observed seemed absurd, even if indeterminacy prevents us from knowing the specific state. This had an anthropocentric corollary: the universe exists only because we (or some entity) observe it. Although such an idea could be used as an argument for the existence of God (who observes all the universe at all times, thus keeping it in existence), such an idea at a human level seemed too outrageous for all but the greatest of egoists. The philosophical pitfalls of deep quantum physics seemed so great that Schrödinger later said he wished he had never met the cat.

Another way that indeterminacy was interpreted was to suggest that there were an infinite number of universes since each quantum transition state should exist. The cat in the box could then be alive in one universe and dead in another.

While scientists have speculated about the various philosophical issues associated with indeterminacy and quantum physics, they are very clear that quantum effects are unnoticeable above the subatomic realm. This has not stopped unscrupulous (or deluded) people from making wacky claims for products ranging from quack medicine to extraterrestrial travel based on "quantum physics."

Evolution, Cellular Biology, and the New Synthesis

The work of biologists in the years before World War I had revealed the existence of genes and chromosomes and had suggested that Mendel's earlier observations contained a possible mechanism for evolutionary change. However, other research showed that the genetic system was more complex. William Bateson (1861–1926) in 1905 demonstrated that some characteristics were blended, rather than segregated, while in the same year Clarence McClung (1870–1946) showed that female mammals have two X chromosomes and males have an X and a Y. This led to the concept of sex-linked characteristics, introduced by Thomas Hunt Morgan in 1910. New tools, borrowed from physics labs, allowed biologists to manipulate nature and to move decisively away from field observation to the probabilistic universe of population studies.

The research on genes and inheritance changed the basis for the continuing discussion of evolutionary theory. Mendelians argued that populations could not have continuous variation and that evolution could happen only through mutation, taking away any need for natural selection as a mechanism. Biometricians (those who followed Darwin more closely) claimed that populations varied around a mean and that the mean could be moved over time. In order to argue that new characteristics would not be swamped and that evolution could happen, biologists needed to think in population terms, which required a statistical approach that borrowed mathematical techniques from the physical sciences. Two such approaches developed, the first through population studies, especially of fruit flies, and the second through a purely mathematical analysis. The result was a rearticu- lation of evolutionary theory, known as the new synthesis.

T.H. Morgan best exemplifies the population approach to understanding evolution. He worked with *drosophila* or fruit flies. These small flies were a perfect experimental subject since they bred quickly, so many generations could be easily traced. In addition, they required little maintenance and were large enough to examine without special equipment. By inducing mutations using heat, chemicals, and X-rays, Morgan traced the distribution of genetic traits. At first, he was skeptical about Mendel's laws of inheritance, but his breeding program tended to confirm Mendel's ideas, although it added to the system in several ways. One change was the discovery of sex-linked characteristics; the "white eye" mutant was almost completely confined to males, thus demonstrating that some changes in the chromosomes were restricted to either the *X* or *Y* chromosome. Further work with Alfred H. Sturtevant (1891–1970) led in 1911 to the first chromosome map, which located the position of five sex-linked genes. By 1916 Morgan was using his findings to argue a genetic theory of natural selection: harmful muta- tions were naturally prevented from spreading (since the individuals with these harmful mutations died out), while beneficial ones gradually took over the population.

Morgan's student, Hermann Joseph Muller (1890–1967), continued the research with *drosophila* to investigate evolutionary change. Muller, who with Sturtevant had helped found a biology club at Columbia University, was fascinated by the concept of genetic change. He carefully studied mutation rates of the fruit flies and concluded that, although there was a certain rate of spontaneous mutation in the genes, the mutations were rarely viable or passed on and, in many cases, were lethal. To demonstrate the physiochemical basis for mutation, he used heat to increase mutation rates (heat increased the chances of random chemical interaction); in 1926 he turned to X-rays, which greatly increased mutation rates. Further, he demonstrated that some of the mutations were inheritable.

This might have been celebrated as a key link between Darwinian evolution and the genetic model of inheritance, but Muller's work was overshadowed by criticism of his politics. Muller was an ardent socialist and even helped publish a communist newspaper. Because of his concern about political suppression in Depression-era United States (the FBI kept tabs on him), he left for Europe in 1932. He was invited by Nikolai Ivanovitch Vavilov (1887–1943), president of the Lenin Academy of the Agricultural Sciences and director of the All-Union Institute of Plant Breeding, to become the senior geneticist at the Institute of Genetics of the Academy of Sciences in Leningrad and later in Moscow. While there, Muller encouraged the Russian school of *drosophila* studies, led by Sergei Chetverikov (1880–1959) and Theodosius Dobzhansky (1900–75). These biologists combined naturalist work with Morgan's lab genetics. They studied fruit fly populations exposed to natural conditions rather than artificially created ones and developed the idea of the gene pool as a reservoir of potential genetic combinations. They maintained strong links with the Columbia research group as well; Dobzhansky traveled to the United States in 1927 to join Morgan's lab for a time.

Muller's time in the Soviet Union was short. He left after only three years, when Vavilov fell from favor around 1937 and genetics as an area of study came under attack. One of Vavilov's most vocal opponents was Trofin Denisovich Lysenko (1898–1976). Lysenko was an agronomist whose reputation was based on a system of winter planting to provide a pea crop before a cotton crop in 1927. He was pictured as a peasant scientist, a practical man who had no time for vague theories or esoteric experiments. His interest was in manipulating the maturation process of seeds, with the aim of bigger plants, higher yields, and shorter growing seasons. He called his system "vernalization," and it involved (among other things) soaking and cooling seeds to promote rapid germination. He claimed he could transform the behavior of plant species through such environmental manipulation, following a neo-Lamarckian form of evolution. Lysenko promoted his system as the solution to all of the Soviet Union's food problems.

Lysenkoism has often been pointed to as a classic case of placing political expediency ahead of good science, but the story is more complex than simply bad science endorsed by a bad political system. Science, and Darwinism in particular, had always had an appeal to communists, who argued that it made the rise of socialism scientifically inevitable. The Soviet Union was still an agrarian society, with a strong interest in biology, but its agricultural system was in deep trouble following the Revolution. The destruction caused by wars, natural disaster, mismanagement, and in some cases deliberate policies of oppression produced famine. The Soviet leadership turned to its leading scientists, such as Vavilov, who said that genetics would help but that it would take time and a great deal of work. Lysenko's system was available immediately and was far more tolerable to a leadership suspicious of intellectuals. The twist on the story was that vernalization did work for a limited number of crops such as peas and corn, but the change in yield was insignificant in a mass system. It did not work at all for other crops such as wheat and, in fact, made things worse in many cases by hurting yields and using up limited resources. Equally damaging to long-term Soviet science was that political support for Lysenko meant the dismantling of rival research programs, especially genetics, which were seen as decadent and Western. Vavilov lost his position and in 1940 was arrested and charged with treason and sabotage. Despite the efforts of the Western scientific community to save him (he was elected a Fellow of the Royal Society in 1942 in order to give him international status), he quickly became an "unperson," his name erased from Soviet records. He was sent to the gulag (Siberian labor camps), where he died in 1943.

What might have been a minor piece of research in agronomics was changed into a national policy by Lysenko's self-promotion and the active support of political leaders who used his work to advance their own agendas. While the politics of a highly centralized state ensured that Lysenkoism with its grave faults caused great damage to Soviet agriculture, it is false logic to conclude that politically motivated support for weak science was confined to totalitarian states, as the cases of cold fusion and missile defense research would later show.

The New Synthesis

The new synthesis gelled with the work of three men: R.A. Fisher (1890–1962), a Cambridge mathematician; J.B.S. Haldane (1892–1964), an Oxford biochemist; and Sewall Wright (1889–1988), an American biologist. These men combined Darwinism, Mendelism, and the statistics of biometry to redefine continuous and discontinuous variation in calculus-like terms, so that the problem of saltation (discontinuous variation) became part of a larger Darwinian continuum. They examined the statistical survival rates of genes and showed that variability was maintained in large populations so that favorable genes could be selected. By thinking in terms of populations rather than individuals, the new synthesis allowed for an integration of the geographical and species concerns of field naturalists with abstract mathematical population genetics. By 1940 scientists had a clear picture of the process of evolution, on both a microscopic and a macroscopic level, opening the door to new debates concerning genetic drift, punctuated equilibrium, and the structure of the gene and genetic material itself.

Studies in population genetics provided much information on the macrobiological investigation of heredity. Chromosome studies had narrowed the site of genetic activity and even mapped the physical location of the control centers or genes. The question now was: How did a gene function? This hinged on a molecular-level examination of nuclear components, and that was a difficult task. Chemical analysis of the fluid in cells showed that each cell contained a soup of different molecules including proteins, enzymes, sugars, and phosphates, among many others. Separating these was a difficult and complex task made possible by the discovery of enzymes. In a sense, enzymes had been used for generations in activities such as fermentation and making bread, but the classification of these organic catalysts as specific chemical entities was first made by Wilhelm Friedrich Kühne (1837–1900), who isolated trypsin, a chemical found in pancreatic juices that aid in digestion. He called these chemicals enzymes, from the Greek *enzumos* meaning leavened. By 1900 dozens of enzymes had been identified, and it was clear that they were the chemical engines of most cell activity. What was not clear was how they were formed and how they worked at a molecular level.

Reactions to Evolution: Social Darwinism and Eugenics

The social reaction to evolution was much different than the earlier revolutions in physics and chemistry, which were often too mathematical and esoteric for popular discussion. Biology, especially human biology, was something that everyone thought they could understand, and the science was about people rather than some inanimate object like a liquid in a test tube or distant planet. Most religious leaders objected to a theory they saw as materialistic and therefore atheistic.

Early public reactions to evolutionary theory focused on the "social Darwinism" of Herbert Spencer rather than Darwin's actual science. The combination of the idea of selection and a social hierarchy based on race that placed Europeans (white people) on top and everyone else seen as lesser creatures contributed to the idea of eugenics. Eugenics was the belief that certain people and certain groups of people had qualities that made them superior or inferior to the social norm and that by the control of reproduction the good qualities could be preserved and enhanced, while the bad qualities could be eliminated. These human breeding concerns show up in Plato's *Republic* and a number of ancient cultures that had rules about the treatment of deformed children and in a world that was predominantly agrarian, breeding people like breeding livestock seemed intuitive. Evolutionary theory was taken by some as providing a scientific justification for the idea. The eugenicists, like the social Darwinists, almost all saw evolution as a ladder of creation with "primitive" forms at the bottom and "advanced" forms at the top. The impetus for eugenics at the end of the nineteenth century was driven by colonialism where white Europeans came to rule over many non-white people around the world, and for the British specifically, the problem of the Irish. The Irish were a colonized people and in Darwin's time many had immigrated to England for work and to avoid the starvation caused by the Great Famine (1845–9). Although Darwin's racism was mild compared to most of his contemporaries, he believed in races and was concerned about population control. In *The Descent of Man* (1871) he argues that a country that started with an equal number of Scots and Irish, would, after a dozen generations, have a population that was five-sixth Irish, but that five-sixths of the property and wealth would belong to the Scots.

One of the founders of modern eugenics was Francis Galton, who invented the term in 1883 combining the Greek roots for "good" and "come into being." He was a pioneer in the use of social statistics and biometry (detailed measurements of the human body). He used these tools to argue that you could identify physical and intellectual differences between people and groups of people. He argued, for example, that the sons of British high court judges (a socially important position that required both education and intellectual power) were likely to also become judges or hold other positions of authority. He concluded that there was some innate, biological reason for successive generations doing so well. This idea appealed to many people, particularly the upper class who could claim that biology and not the luck of some distant ancestor accounted for their success.

Sybil Gotto (1885–1955) and Galton founded the Eugenics Education Society in 1907 and the society began publishing *The Eugenics Review* in 1909. Gotto was an anti-poverty activist and promoted sex education. She hoped to reduce poverty in part by reducing the number of children born to poor families and preventing people of low intelligence from having children they could not support. In a time of mass poverty in the growing industrial cities, these ideas found widespread support and the society attracted many leading figures in British society including members of the clergy and the aristocracy. Countries around the world introduced eugenics laws at the beginning of the twentieth century and many of those laws included involuntary sterilization.

Eugenics has always had two aspects, sometimes referred to as positive eugenics and negative eugenics. The positive form encouraged "good" people to have more children and worked to prevent those considered less good from having children through education and voluntary birth control. The negative form of eugenics was the use of involuntary birth control (usually sterilization) and on occasion murder to prevent those people deemed unfit from having children. Historically, this distinction was meaningless. The atrocities of the Nazis' eugenics program were built on the same ideas as those proposed by the Eugenics Education Society. All were equally flawed. Although it seems intuitive to think that humans can be bred to enhance certain characteristics, the biology is much more complex. The simplest problem, which Galton himself noted, was regression toward the mean. For example, two exceptionally tall parents are more likely to have children who are closer to the average height of the population and are less likely to be even taller than their parents. When it comes to something really complex such as intelligence, there is no genetic formula for smartness and factors such as pre- and postnatal nutrition or family wealth play major roles in cognitive development. Further, the emerging study of epigenetics (the study of genetic changes that do not alter DNA sequences) has made clear that your DNA is not your destiny.

Reactions to Evolution: Darwinism, Education, and Religion

Somewhat ironically, the most famous early twentieth-century public dispute about evolution, the so-called Scopes Monkey Trial, occurred at a time when many biologists themselves were unsure about Darwinism. The 1925 trial in Dayton, Tennessee, must be set against the background of social turmoil in the United States, especially in the economically depressed South, and the disruption caused by the war and postwar years. Many people linked science to outsiders or foreigners, and feelings about scientists were often heavily influenced by anti-Semitism or fear of atheism, which was associated with the materialism of science.

The Tennessee House Bill 185 specifically banned the teaching of evolution in any school that received public money from the state. George Rappalyea, a geologist and mine owner, asked the American Civil Liberties Union (ACLU) to finance a court challenge; when they agreed, he asked his friend John Scopes, a young science and sports teacher, if he would be willing to be arrested for the test case. The trial was a media sensation, bringing together two famous lawyers, Clarence Darrow (for the defense) and William Jennings Bryan (for the prosecution). Reporters from across the country and around the world arrived in Dayton hoping to see a major clash. While the trial was a circus (both Darrow and Bryan were made honorary colonels in the state militia, for example), it was not quite the battle of science and religion that some had hoped for. The judge ruled that evidence for evolution could not be admitted, preventing the defense from introducing expert testimony from scientists. But in one of the strangest twists in legal history, Darrow called Bryan to testify as an expert witness on Genesis. In the end Scopes was found guilty and fined $100, which was actually the result desired by the ACLU, since they could only take the case to a higher court and have the legislation overturned if there was a conviction. Unfortunately, the ruling was overturned on appeal due to a technicality, making a Supreme Court decision impossible and allowing the original legislation to stand. A number of other states enacted similar laws against the teaching of evolution, and it was not until 1968 that the Supreme Court ruled that all such specifically anti-evolution statutes were unconstitutional since they violated the Establishment Clause of the First Amendment regarding the separation of church and state. This ruling was tested in 2005 in the case of *Kitzmiller v. Dover Area School District*. The school board changed the biology curriculum to include "intelligent design" (a theory that biological complexity can only arise from an intelligent designer, not evolution). The court ruled against the school district, saying that intelligent design was a religious, not a scientific idea.

The struggle over evolution and education continues today. A number of countries restrict the teaching of evolution or include religious material in the biology curriculum of public schools. Very few countries require private schools, especially religious private schools, to include evolution (or even science generally) in their curriculum.

Science and the State: The Atomic Bomb

In Germany and Italy, the Fascists used genetics and evolutionary theory to impose their racist ideology. The Nazis ordered a purification of "Aryan blood." In the midst of economic depression, deprivation, and anger, Hitler and Mussolini found many willing supporters for their radical political ideals. German scientists, in particular, found themselves in a difficult position. To continue their work, they had to accept the imposition of political control and in many cases even join the Nazi Party. Some did this willingly, many out of necessity. Others, particularly after Adolf Hitler seized power in 1933, chose to leave Germany, Italy, and Austria to escape fascism. Some, like Albert Einstein who left Germany in 1932, did not have much choice because they were Jewish or had too close associations with groups rejected or outlawed by the new rulers. They could stay and give up their scientific careers and perhaps their lives, or they could leave.

Among the escaped scientists were physicists who brought with them the fear that their German colleagues were developing a super weapon. They recognized how the advances made in radioactivity research by people such as the Curies and Rutherford and the theoretical insight into the relationship between mass and energy articulated by Einstein suggested the potential power of materials such as uranium to produce a super bomb.

10.1 LISE MEITNER

The secrets of radioactivity were still being unraveled after World War I, since the process turned out to be rather difficult for a number of reasons. One of the most practical was the lack of material to work with, since radioactive elements were rare and hard to refine. The second was contamination. Radiation affected laboratory equipment (and the health of the physicists themselves), and even though the amounts of material were small and precious, they were spilled, smeared, and dissipated around the labs until the labs were so contaminated that researchers had to abandon them and move to fresh space. Finally, radioactive materials were hard to work with because they didn't sit still. As they radiated, they literally turned into new substances.

One of the most fruitful research programs involved bombarding radioactive substances with neutrons. Enrico Fermi (1901–54), following work done by Jean Joliot and Irene Joliot-Curie, demonstrated that radioactive isotopes of many elements could be produced by neutron bombardment. In turn,

Lise Meitner (1878–1968) with her research partner Otto Hahn (1879–1968), who had been working on problems in radioactivity since 1907, turned to the products of uranium bombardment around 1936. Meitner was a remarkable scientist, one of the first women to graduate from an Austrian university after they were opened to women. She traveled to Berlin and, against the advice of many, studied advanced physics, earning her doctorate in 1906. She became Germany's first female physics professor in 1926 and the first woman to receive a salary as a researcher at the Kaiser Wilhelm Institute.

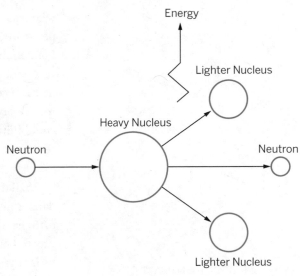

NUCLEAR FISSION

10.2 FISSION

In 1933, when Hitler took power, Meitner was protected because she was Austrian, but in 1938 Germany annexed Austria, and Meitner came under German law. As a Jew, she was in danger, so she fled to Sweden. Hahn, who continued working with Fritz Strassmann (1902–80) on the problem of the creation of radioactive isotopes, corresponded regularly with her. He and Strassmann produced a series of experiments in which they bombarded uranium with neutrons, unexpectedly producing barium, a much lighter element. When Hahn sent Meitner the strange results, she followed the "liquid-drop" model of the nucleus suggested by Niels Bohr to determine that the uranium nucleus must have split apart and this explained the formation of the lighter element, barium. In Bohr's model, a neutron hitting the nucleus of a heavy element might do one of three things: it might get stuck, making the nucleus heavier by a neutron; it might knock off a bit of the nucleus, releasing a few protons and neutrons; or it might cause the nucleus to break apart, converting a small amount of matter to energy. Meitner called the splitting *fission*. She sent a letter with her conclusion to the periodical *Nature* early in 1939. (See figure 10.2.)

Bohr was at a conference in the United States when he heard about Meitner's finding. He rushed to tell other physicists about the discovery, and it caused a sensation. Several went back to their labs and replicated the discovery, confirming the work of Meitner, Hahn, and Strassmann. To Leo Szilard (1898–1964), however, the discovery opened up the horrific possibility of a nuclear bomb. Szilard,

UNMODERATED CHAIN REACTION

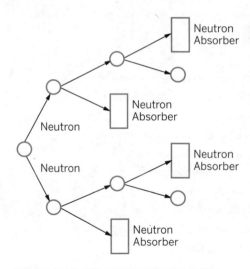

CONTROLLED FISSION CHAIN REACTION

10.3 CHAIN REACTIONS

who was Hungarian, left Germany in 1933 to avoid persecution. That same year, while walking around London, he conceived the idea of a neutron chain reaction. The idea was a remarkably simple one. Since a neutron could strike an atomic nucleus and cause the release of one or more neutrons, these released neutrons could initiate a potentially continuous number of reactions, releasing an awe-inspiring amount of energy. (See figure 10.3.)

Szilard regarded this possible process to be so dangerous that, when he got a patent for the idea in 1936, he assigned it to the British Admiralty, the only way to both register the patent and keep it secret. When he learned about the fission of uranium, he recognized that a practical path to a neutron chain reaction was now available, and the results could be devastating.

World War II began in 1939, pitting the Axis powers of Germany, Italy, and Japan against the Allied forces of Britain (and the Commonwealth), France, and Belgium. The power of the German blitzkrieg overwhelmed Allied forces and Belgium surrendered in 1940, putting the Belgian Congo, then the greatest known source of uranium, in German hands. Werner Heisenberg sent a secret paper entitled "On the Possibility of Technical Energy Production from Uranium Splitting" to the German Army Weapons Bureau in early 1940. The threat of a German atomic bomb seemed very real. Suddenly, the ingredients for a super weapon were in Hitler's hands. Germany certainly had the industrial capability to create such a weapon, and, even though a number of the best and brightest physicists had fled Germany, there were still powerful minds such as Hahn and Strassmann available to work on such a project.

Szilard's early efforts to interest the American government in the idea of an atomic bomb had little apparent effect, so in 1941 he persuaded Einstein (now living in the United States) to write directly to President Franklin D. Roosevelt. Although this letter was very influential in the decision to proceed with the project, the military also consulted with other scientists such as Enrico Fermi, Neils Bohr, and John von Neumann about the potential of atomic power and weapons as they made their decision to go forward with the bomb.

One of the key figures who advised President Roosevelt on science policy was Vannevar Bush (1890–1974). He convinced Roosevelt to create the Office of Scientific Research and Development (OSRD), a federal research body, and he served as its head along with James Conant (1893–1978), president of Harvard University. One of the OSRD's main interests was nuclear power. Bush was an electrical engineer with experience in military research during World War I; he was also a vice president of the Massachusetts Institute of Technology (MIT) and head of the Carnegie Institution. He planned to use a network of universities to do research rather than expand or establish new federal laboratories. The universities were happy to take on the work since the funding provided was on a scale unseen since the days of chemical warfare. Under this system of federal money directed to private researchers, many of the most important American research centers were established, including the Jet Propulsion Laboratory (JPL) at the California Institute of Technology (Caltech), the Radiation Laboratories under the auspices of the University of California and MIT, and the Metallurgical Laboratory at the University of Chicago. Federal money for research (not including the Manhattan Project) rose from $74 million in 1940 to $1.59 billion by the end of the war.

The Manhattan Project

In October 1941 Roosevelt was briefed on the potential of building a nuclear weapon. He authorized this research on December 6, 1941, the day before the Japanese attack on Pearl Harbor. The army's initial part of the top-secret project was called the Manhattan Engineering District so as to hide its true nature (and location). It brought together some of the most powerful minds in science in a race to beat the Germans to a super bomb. Better known simply as the Manhattan Project, it changed the course of world history.

The first steps toward a weapon were to confirm the possibility of a sustained chain reaction by fission and to evaluate the technical problems associated with the production of the necessary materials. The fission problem was undertaken by a team headed by Italian émigré physicist Enrico Fermi, who had already won a Nobel Prize for his work on radioactive elements. When he traveled to Sweden in 1938 to receive the prize, he was criticized in Italy for failing to wear a Fascist uniform or give the Fascist salute. He and his family took the opportunity of their trip abroad to escape and never returned to Italy. Fermi began his work on the Manhattan Project at the University of Chicago. He built a small reactor, called an atomic pile, under the bleachers at the west end of Stagg Field in a space that had previously been squash courts. It was constructed of blocks of graphite (a kind of carbon that absorbed neutrons), uranium, and uranium oxide. Control rods made of cadmium, designed to limit the rate of fission, were inserted in holes through the blocks. On December 2, 1942, at 3:25 in the afternoon, Fermi's team slowly pulled out the control rods and a controlled self-sustaining fission reaction began. (See figure 10.3.) This was the official start of the "atomic age."

Arthur Compton (1892–1962), a member of the committee investigating the possibility of building a nuclear bomb, was present at the reactor test. He called James B. Conant at Harvard to pass on, in code, the news of the success in a now famous telephone exchange:

> "The Italian navigator has landed in the New World," said Compton.
> "How were the natives?" asked Conant.
> "Very friendly."[3]

With a number of the theoretical questions answered by Fermi's atomic pile, the next stage was to produce an uncontrolled chain reaction. This was a big project in terms of cost, scope, and originality. Brigadier-General Leslie Groves was selected as the military head of the project. Trained as an engineer, he had experience organizing large construction works, including the building of the Pentagon. He chose Robert Oppenheimer (1904–67) as the scientific director. Ultimately, they would be in charge of hundreds of scientists and thousands of military and civilian workers.

Two major hurdles had to be overcome. The first was the uranium itself. At the start of the war, the amount of refined uranium in the world could be measured in grams, but tons of uranium ore would be needed for the Manhattan Project. With the best-known sources in Europe and the Belgian Congo under the control of Germany, new mines were needed. A massive search for ore was undertaken. Sources had to be secure and in friendly territory. Fortunately, large resources were found in northern Canada, which became one of the main

---

3. Arthur Compton, *Atomic Quest* (New York: Oxford University Press, 1956), 144.

suppliers of raw ore. Indigenous workers who had carried out the radioactive material reported disturbing trends of chronic illnesses and cancers in later years.

The second problem was a sticky technical issue about the kind of material needed. Neils Bohr had pointed out that the isotope U_{235} would sustain fission far better than U_{238}. While both occurred naturally, only about 0.7 per cent of uranium atoms were U_{235}. Separating the two was an enormous job because they were chemically identical and only about 1 per cent different in mass. Three methods of separation were tried: magnetic separation, gaseous diffusion, and gas centrifuge. Magnetic separation looked promising. Uranium tetrachloride in gas form was passed across a strong magnetic field. The heavier U_{238} was deflected less and thus separated from the desired U_{235}. This system was created by Ernest O. Lawrence (1901–58), one of the inventors of the cyclotron, at the University of California at Berkeley, but after millions of dollars were spent on the factory, it failed to produce the needed quantities. The gas centrifuge worked in experimental operation but could not be scaled up to industrial capacity. That left gaseous diffusion. As uranium hexafluoride gas was passed through porous clay filters, the lighter U_{235} passed more easily; after repeated filtration, uranium of the required purity was obtained. The massive Oak Ridge plant in Tennessee used this method to produce much of the material for the uranium bomb.

With the uranium supply issue resolved, a new material problem was created when Glenn Seaborg (1912–99) suggested that plutonium offered even better fission properties than U_{235}. The best isotope of the recently discovered element Pu_{239} was produced by placing U_{238} in a reactor and letting it pick up neutrons. Uranium was converted to plutonium, and the plutonium was enriched by having neutrons added to make fissionable material. Such a reactor was called a "breeder reactor." Plutonium was therefore added to the production system, and Seaborg was put in charge of producing it for the project.

The scientific team was assembled at Los Alamos, New Mexico, to work out the scientific and engineering difficulties involved in creating an explosive reaction. At the heart of the problem was getting the mass of material at the center of the device to go from no sustained fission to fission at a specific moment. They achieved this by creating an implosion with conventional explosives that compressed the fissionable material and set off the uncontrolled chain reaction. The effort to bring together all the elements took several years, but finally the Trinity Test, of a bomb called Gadget, took place at Alamogordo, New Mexico, on July 16, 1945. The Trinity Test exploded at about 5:30 a.m. Enrico Fermi, an eye witness, estimated the power was about equal to 10,000 tons of TNT.

CONNECTIONS

Science and
Fascism

On April 7, 1933, the German government passed a law called the *Gesetz zur Wiederherstellung des Berufsbeamtentums*, the Law for the Restoration of the Civil Service. This law targeted "non-Aryan" civil servants and forced them out of their jobs, including those scientists who worked at German universities and research centers such as the Kaiser Wilhelm Institutes. Although Max Planck made a personal appeal to Adolf Hitler to reverse this decision, Jewish scientists were forced to leave their jobs and to flee the country. A further group of scientists fled Italy when Mussolini came to power and Austria after the Nazi Anschluss in 1938.

Fascist ideology was based on racism, but it also attacked the idea of the universality of science. Hitler explicitly attacked what he called "Jewish science," including the theory of relativity, quantum mechanics, and uncertainty theory. He placed Jewish science in the same category as jazz music and modern art as degenerate forms of culture. This policy took a terrible toll in human terms, and it also transferred the center of modern physics research from Germany to the United States and Britain.

Albert Einstein left for Britain in 1932, before the law had been passed, and was instrumental in publicizing the hardship of scientists displaced by fascist policies. The economist William

Oppenheimer quoted the *Bhagavad-Gita*, saying "now I am become Death, the destroyer of worlds...."[4]

The explosion presented both a scientific triumph and an ethical dilemma. Germany had surrendered on May 7, 1945. The race against the Germans was over. For many of the scientists, especially the Europeans displaced by the Fascists, that meant the end of the need for a super bomb. What few realized was that as early as 1943, Groves had already been looking at the use of the weapon in the Pacific theater. The construction and delivery of the weapons were pushed ahead. In the end, three bombs were constructed: Little Boy, Fat Man, and Bomb #4.

....

4. Robert Oppenheimer, quoting the *Bhagavad-Gita*, in Ferenc Morton Szasz, *The Day the Sun Rose Twice* (Albuquerque: University of New Mexico Press, 1984), 89.

Beveridge set up an Academic Assistance Council in the United Kingdom, aimed at rescuing Jewish and politically vulnerable academics. Many prominent academics quickly backed this organization – J.B.S. Haldane, John Maynard Keynes, Ernest Rutherford, G.M. Trevelyan, and A.E. Housman, to name a few. This organization eventually helped over 1,500 academics escape Germany and Austria, in order to continue their work in Britain and other free countries.

The list of physicists, mathematicians, chemists, and other scientists who escaped from fascist Europe was remarkable. Physicist Rudolf Peierls (1907–95) went on to work with the British team connected to the Manhattan Project. Nuclear physicist Hans Albrecht Bethe (1906–2005) was fired from his position in the University of Tübingen and joined Peierls in 1933. Leo Szilard fled to Britain in 1933 as did Max Born (1882–1970) who went on to write a best-selling science book, *The Restless Universe*. Lise Meitner escaped in 1938, traveling first to the Netherlands and then to Sweden. Enrico Fermi escaped fascist Italy in 1938 when he traveled to Stockholm to receive the Nobel Prize in Physics and never returned. Many of the scientists moved to the United States and were instrumental to the Manhattan Project. A Stanford economist, Petra Moser, has estimated that US patents increased by 31 per cent in fields common among Jewish scientists who fled Nazi Germany for America. Of the scientists who went to Britain in the years before and at the beginning of the war, many went on to win Nobel Prizes, to receive knighthoods, and over 100 became members of the Royal Society or British Academy.

Historians have speculated that German racial policies, besides being inhumane in and of themselves, used up massive resources and crippled German scientific work. Refugee scientists helped give the Allies a scientific and technological advantage over the Nazis and contributed to establishing Big Science in the Western world.

Leo Szilard was horrified that the project was still continuing and once again went to Albert Einstein for a letter of introduction to Roosevelt. Szilard wanted to persuade the president not to use the weapons. He was concerned about both the devastation they would cause and the resulting arms race that might destroy the planet. Before the meeting could be arranged, Roosevelt died on April 12, 1945. Harry S. Truman became president, and among his briefings on assuming the office he learned about the Manhattan Project. Szilard could not meet Truman, meeting instead with James Byrnes, Truman's secretary of state. Byrnes had already advised Truman to use the bomb as soon as possible, and he rejected Szilard's concerns.

Truman's advisors concurred with the military that the bombs should be used without warning on Japan. Further, targets were to be chosen so that the effects of the blasts could be studied. Truman faced a hard decision. Although the

Japanese were losing badly, they were offering stiff resistance and publicly vowed to fight to the death. Military planners suggested that as many as 1 million casualties might be expected from an invasion of mainland Japan, whereas many fewer would die from the bomb attacks. There was still anger over the sneak attack on Pearl Harbor, and to further complicate the situation the defeat of Germany had freed the Soviet military to join the battle against Japan. Truman wanted to avoid the division of Japan following the precedent of the partitioning of Germany if the Soviet Union were to participate in the war.

Therefore, Truman authorized the use of the atomic bombs on July 21, 1945. On July 26 the Allies released the Potsdam Declaration calling on the Japanese to surrender unconditionally or face "prompt and utter destruction." Two days later the Japanese government rejected the call for surrender. Preparations were made for dropping the bombs. With complete air superiority, there was no interference on August 6 as the *Enola Gay* flew to its target and dropped Little Boy on Hiroshima. On August 9 Fat Man destroyed Nagasaki. Japan surrendered unconditionally on August 14. More than 200,000 people were killed in the two nuclear blasts, the vast majority of them civilians. Many more were injured, some from radiation burns and poisoning.

The effort to build the bombs cost about $1.8 billion. This can be compared to the $5.4 billion the United States spent on tanks and the $2.6 billion spent on all other explosives. Although building the nuclear bombs was not exactly a bargain, given the results many felt it was a justifiable expense. It remains a point of debate, however, whether the race for the bomb had really been necessary at all. At the end of the war there was little evidence that nuclear research was being undertaken in Germany, although some recently declassified documents suggest that key German scientists might have been able to construct such a weapon. The documents emerged after the war during Operation "Alsos," in which the Allies debriefed ten of Germany's top scientists, including Otto Hahn and Werner Heisenberg. Their discussion of the bombing of Hiroshima and Nagasaki suggests that they already understood the technical details as well as the basic principles behind the bomb. This changes the question from "Were the Germans working on an atomic bomb?" to "Why didn't they build one?"

National Security and Science Policy

In some ways, the Trinity Test let the nuclear genie out of the bottle, but in the wake of the bombing of Japan, as Szilard predicted, there followed an arms race that put the safety of all life on the planet at risk. It raises, even more than the involvement of scientists in chemical warfare, the issue of what role scientists have to play in military activity. Were the scientists responsible for the bomb, or did that responsibility lie with the politicians and military leaders? To what extent should national security determine science policy?

The postwar years provided a simple answer for many. Where nuclear weapons and national security were concerned, scientists were expected to work for the state. Three things were needed to manufacture a nuclear weapon from scratch: access to uranium, an industrial infrastructure capable of manufacturing the components of the weapon, and intellectual resources sufficient to manage the first two elements. At the end of the war, four nations had the capacity to build nuclear weapons: the United States, Canada, Britain, and the Soviet Union. The United States had the bomb already, while Canada, as one of the lesser powers, did not pursue an independent nuclear program, preferring to disarm and not spend the money. The Soviet Union began its work on a weapon shortly after the war, and on August 29, 1949, at Semipalatinsk in Kazakhstan, detonated its first atomic bomb, Joe 1, with about a 10- to 20-kiloton blast. British scientists, some of whom had taken part in the Manhattan Project, were pressed into nuclear work. Although badly battered by the war, Britain put together its own nuclear program and tested its first weapon in 1952.

Even before the end of the war, the alliance between the Western powers and the Soviet Union was strained. Both sides saw the other as the next enemy. General George S. Patton even secretly discussed taking "his boys" on to Moscow after he defeated the Germans. So it was not surprising that, unlike during the years following World War I, the scientific teams of World War II were not disbanded and sent back to the groves of academe. Rather, in this charged environment, which has become known as the Cold War, the next step in destructive weapons was taken. As early as 1938 Hans Albrecht Bethe had studied thermonuclear reactions involving light elements in order to understand how the Sun (and all stars) functioned. He concluded that under the right conditions hydrogen would undergo fusion. At the great temperature and gravitational pressure inside a star, hydrogen atoms were squeezed together (or fused) to form helium. When that happened, a small amount of mass was converted to energy. The basic principle looked simple: $_1H^2 + _1H^2 \rightarrow _2He^4$. (See figure 10.4.)

The energy of the stars comes from the loss of mass that accompanies the transformation, following Einstein's relationship of $E = mc^2$. About 0.63 per cent of the mass of the original atoms is lost by conversion to energy. While this seems

10.4 THE FUSION OF HYDROGEN TO HELIUM

tiny, it is huge compared to the energy derived from fission, in which the splitting of a uranium atom releases only 0.056 per cent of the mass as energy. In other words, fusion released more than ten times as much energy as fission. Hence the power of the Sun, which converts about 650,000,000 tons of hydrogen to helium every second.

Edward Teller (1908–2003) had also been thinking about the power of fusion in the early 1940s. Following the invention of the fission bomb, he saw a way to create conditions for a fusion bomb. There is still a great deal of debate over the creation of the first fusion bomb, or H-bomb. A priority fight between Stanislaw M. Ulam (1909–84) and Teller as well as the secrecy surrounding the nuclear arms race have clouded the exact history, but it is likely that the theoretical work was done by Teller and others at Los Alamos, and the initial design of the device was done by Richard L. Garwin (1928–), a young physicist on a research visit to the facility. The basic structure of the hydrogen bomb was a fission bomb surrounded by, or next to, a high hydrogen package. The fission bomb exploded in such a way that it reached the ignition point for a fusion reaction. Teller championed the development of the weapon, and the first test bomb was ready in 1952. It stood two stories high, and when it was detonated, it vaporized the Pacific island of Elugelab. Its power was equal to 10.4 million tons of high explosive, or about 700 times the power of the atomic bomb dropped on Hiroshima.

Unlike the fission bomb, there was no theoretical limit to the destructive power of a hydrogen bomb. Scientists conceived of weapons powerful enough to blow huge holes in the atmosphere, flatten entire countries, or create gigantic tidal waves. The measurement system for destructive power jumped an order of magnitude from kilotons to megatons. In addition to the raw destructive power of the fusion bomb, there was also massive radiation danger. The larger weapons released far more radioactive materials such as strontium-90 and cesium-137, which they blew high into the atmosphere from where they rained down on a wide area.

A number of scientists opposed the development of these super bombs. Robert Oppenheimer was stripped of his security clearance and effectively prevented from doing further research for the military because of his opposition to the development of bigger weapons. Beginning in 1947 the *Bulletin of Atomic Scientists* placed a doomsday clock on its cover. The first clock was set at seven minutes to midnight. In 1953, after both the United States and the Soviet Union tested fusion weapons, it was set to two minutes to midnight. In 1955 Bertrand Russell and Albert Einstein issued a manifesto calling on all governments to find peaceful means to resolve conflicts and to give up nuclear arms. It was signed by Max Born, Percy Bridgman, Leopold Infeld, Frederic Joliot-Curie, Hermann Muller, Linus Pauling, Cecil Powell, Joseph Rotblat, and Hideki Yukawa. Although the war effort had forced scientists to choose national interests over the international community of science, the profile of these scientists indicates a countermove to recreate an international republic of science. From the end of World War II, these competing loyalties vied with one another, with national pride and security (and funding sources) pulling against a belief in the universal nature of scientific understanding. Einstein said, "I know not with what weapons World III will be fought, but World War IV will be fought with sticks and stones."[5]

Discovering DNA

One of the signatories of this manifesto, the geneticist Hermann Muller, studied mutations in genes and was particularly aware of the dangers of radiation from his biological work. In one of the ironies of modern science, the science of destruction and the science of life drew on material from the same source. X-rays and radioactive tracers played a key role in taking the work of the geneticists into the heart of the cell. Quietly, almost without notice by the public or the politicians, whose attention was on physics, the quest for the control mechanism in inheritance continued. Using many of the resources and techniques from the physics lab, genetics and cell biology gained momentum in the aftermath of World War II.

The area of investigation shifted from tracing population inheritance patterns to understanding the structure of the genetic material itself. Biologists were influenced by the understanding of nature at an atomic and subatomic level that physics provided. They increasingly sought to unravel genetics at the chromosomal and molecular level. The first step was to identify the substance within the cells that contained genetic information.

5. Alice Calaprice, *The New Quotable Einstein* (Princeton, NJ: Princeton University Press, 2005), 173.

At about the same time that Mendel was working on pea plants, the German scientist Friedrich Miescher (1844–95) was analyzing pus from discarded hospital bandages, hoping to find a cure for infections that frequently occurred in bandaged wounds. He tested the material he had isolated and discovered that some of it was not a protein. This was a puzzle: What was this substance doing in a mass of proteins? He found that the material, which he called nuclein (later nucleic acid), contained a high level of phosphorous and concluded that it was part of the nucleus of a cell and acted as a storage space for phosphorus, which the body needs in small quantities. In the 1880s August Weismann, while investigating chromosomes, discovered that this same nucleic acid was present in them. By the time Morgan's lab at Columbia University began its *drosophila* experiments, it was accepted that genetic information was passed on through something in the chromosomes, either the protein or this nucleic acid.

In 1928 Fred Griffith (1881–1941), a British medical officer, was studying the bacteria responsible for a pneumonia epidemic and found that the pneumococcus bacteria occurred in two forms. One form (*S*) was smooth and highly infective; the other (*R*) was rough and harmless. Because both were found in patients, he wondered what the relationship between the two might be. Griffith injected mice with harmless living *R* cells and heat-killed *S* cells. The mice died, and living cells of both types were found in their bodies. He concluded that some substance from the *S* cells was transferred to the *R* cells, changing them into virulent *S*-type cells. If it could be isolated, this would be the control substance.

In 1944 the team of Oswald Theodore Avery (1877–1955), Colin MacLeod (1909–72), and Maclyn McCarty (1911–2005), at the Rockefeller Institute in New York, reported that they had isolated Griffith's transforming material and that it was a nucleic acid – specifically, deoxyribonucleic acid, or DNA. Although not everyone was convinced, the large molecule came under increased study. The Phage School (named for the phage viruses they studied) used radioactive tracers to follow molecular events in phage infection; in 1952 A.D. Hershey (1908–97) and Martha Chase (1930–2003) discovered that phages leave their protein coats behind and infect bacterial cells with their DNA. There is much debate as to whether the Avery/MacLeod or the Hershey/Chase experiments constitute the moment of the identification of DNA as the genetic material. In many ways, the answer depends on which professional subdiscipline (molecular biology or bacteriology) is considered more important, not which one supplied the definitive answer. The two findings together, however, pointed all interested geneticists toward the structure of the DNA molecule as the key to understanding inheritance.

The work of Hershey and Chase had more immediate impact than that of Avery and MacLeod because the chemical composition of DNA had been determined. In 1950 Erwin Chargaff (1929–92) made a major step in unraveling the complex nature of DNA. He established that the molecule contained four types of nitrogenous bases, which existed in a one-to-one ratio of adenine to thymine and guanine to cytosine. This held true for all the different samples from a range of organisms. These bases could follow each other in any arbitrary order on the polynucleotide chain, so there was a possibility that the order of the bases somehow affected inheritance. This discovery was one of the key components needed to create a model of DNA.

The race to discover the structure of DNA was on. It has become one of the iconic case studies in the history of science, raising questions about how scientists actually work, the way credit is awarded, and what effect cultural expectations, both within the scientific community and in the larger society, have on the practice of science. Historians have been able to ask these questions about these events because many of the people involved reported what happened. Particularly telling in this regard was the frank account of his own work written by James Watson in *The Double Helix* (1968) and an alternative view of events found in Anne Sayre's study, *Rosalind Franklin and DNA* (1975).

A number of laboratories were working on DNA. Linus Pauling (1901–94) in the United States turned his attention to the structure, while at the University of London Rosalind Franklin (1920–58), in conjunction with Maurice Wilkins (1916–2004), used X-ray crystallography to analyze DNA. Franklin did her scientific training at Newnham College, Cambridge, obtaining her MA in physical chemistry in 1947. After a series of research positions, including the Centre national de la recherche scientifique, she obtained a fellowship at King's College, London. In 1951 Franklin produced extremely fine X-ray diffraction photographs of what was called the B form of DNA, but she did not immediately see the structure, because it was only one of a number of different images. She was also in an unpleasant situation at London, as Wilkins treated her like a technician rather than a colleague, and the male-only world of the university and the laboratory excluded her from the informal networks so important for both support and the contacts that often underlie scientific work.

Into this situation came Francis Crick (1916–2004) and James Watson (1928–). Crick had been employed in war work as a physicist but was now working on his PhD at Cambridge in biophysics, having read and been influenced by Schrödinger's book *What Is Life? The Physical Aspects of the Living Cell*. He met

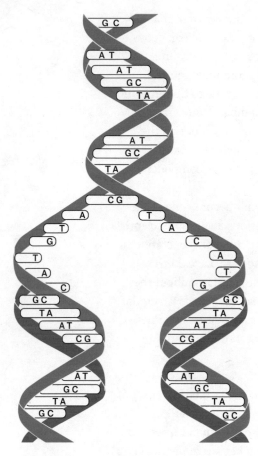

10.5 THE DOUBLE HELIX

James Watson, an American who had already earned his PhD at the age of 22 and who had done work as a phage geneticist. Although both men were supposed to be working on other projects – Crick on his thesis and Watson on viruses – they decided to try to find the structure of DNA. They set out to build a model that would agree with the X-ray diffraction data and account both for autocatalysis (DNA splitting in half) and heterocatalysis (transferring information to create proteins and other cells, as took place in reproduction).

It was clear to Watson and Crick that the molecule was a long-chain polymer with a constant diameter. They already had the basic chemistry of bases and sugars, and they knew that any arrangement of nucleotides in the DNA molecule had to account for the regularity of the molecule's structure as well as its chemical stability. It also had to account for how the molecule could replicate itself faithfully. Three factors influenced their thinking. First, unlike bases seemed to attract each other. Second, Chargaff had shown that the ratio of bases was 1:1. Finally, Pauling, whose model building inspired theirs, introduced the idea of the helix. While an early attempt went badly wrong and they were warned by their superiors to concentrate on their own work, Watson and Crick persisted. One major breakthrough came when Maurice Wilkins showed them, without Franklin's knowledge or permission, her crystallographic photographs of DNA. The image revealed that the molecule had to be a double helix; Watson and Crick constructed a model that looked like a twisted ladder made of two spines with the base pairs (adenine, thymine, guanine, and cytosine) arranged as rungs. (See figure 10.5.) It accounted for genetic, biochemical, and structural characteristics and explained auto- and heterocatalysis. Equally, it predicted the mechanism of storage of genetic information in the order of the base pairs. On April 2, 1953, Watson and Crick published their model in the journal *Nature*. The brief paper, only a

10.6 WATSON AND CRICK AND THE DOUBLE HELIX

Source: A. Barrington Brown / Science Source.

single page, was a sensation. Their model was confirmed, and the cellular control system was revealed. For this work they won the Nobel Prize for Physiology or Medicine in 1962, sharing the honor with Wilkins. (See figure 10.6 of Watson and Crick with their model of DNA.) Franklin, who had spent her career working with X-rays, had already died of cancer and so was not eligible to share the prize.

There is no doubt that Watson and Crick made a brilliant discovery, but their methods seem to run contrary to what is commonly taught as good scientific practice. They did no primary research but rather sought information from others, with little to share in return. Their access to, and use of, Franklin's work seems underhanded and questionable, and they did not acknowledge her contribution in their initial publication. On the one hand, scientific information was regarded as "public" in the sense that it was (and still to a large extent is) expected that work and results would be talked about informally over dinner or beer, as well as being presented formally at conferences and published. It was also accepted that others would make use of that material, regardless of its formal or informal source. On the other hand, professional success was based in part on priority and the acknowledged importance of work. If scientists used other people's work, they were expected to acknowledge their sources. Watson and Crick ignored part of the code of gentlemanly behavior implicit in the ideology of science dating back to the scientific revolution. The race for the structure of DNA suggests that in science "anything goes" (as philosopher Paul Feyerabend claimed) and that its ideology has come to resemble more closely some Darwinian struggle for existence than the gentlemanly witnessing espoused by Robert Boyle. Should scientific research be constrained by rules of polite conduct, or does the pursuit of knowledge override such limitations?

Conclusion

The utility and power of science was increasingly recognized by governments in the postwar years, and the organization of science was transformed as a result. With the start of the Cold War, governments did not dismiss their teams of scientists as they had after World War I but instead chose to support scientific research and direct it to politically determined objectives. All the industrial nations and many of the less technically advanced countries recognized the necessity of having and developing scientific knowledge and integrating national interests with research interests. After World War II, almost all-important scientific breakthroughs came from scientific teams rather than individual scientists. These teams were increasingly funded by major research institutions or by federal governments through grants and contracts. While the emergence of "Big Science" had its roots in the Chemists' War, it was really the Physicists' War that cemented the power of large laboratories and established the massive infrastructure needed to support them.

In disciplinary terms, physics replaced chemistry as the premier science and became the focus of increasing public, governmental, educational, and industrial attention. Biology gained by its connection to physics but remained a distant second in terms of funding. While the military forces of the world powers continued to regard science and scientists with some suspicion, the effectiveness of science-backed military activity was demonstrated first by Germany's blitzkrieg and more spectacularly by the first belligerent use of nuclear weapons. The utility of science after the atomic bomb was too powerful to allow scientists to slip back into their quiet academic laboratories as they had largely done at the end of World War I. The new way to do science had been proven.

"Big Science" with big funding, big laboratories, and larger and larger teams of scientists working on research agendas established by national governments was the way of the future. The United States and the Soviet Union, staring at each other across an ideological chasm, mustered their scientific forces. This marshaling of science produced the new arms race and its corollary, the race to space. The place of science and the practice of research were permanently changed.

Essay Questions

1. What was the "new synthesis" in biology?

2. What two ideas were needed to make an atomic bomb possible?

3. Did the discovery of the structure of DNA circumvent traditional methods of scientific research?

4. Why was indeterminacy necessary for modern physics?

5. In what ways did the eugenics movement seek to "improve" the human race?

CHAPTER TIMELINE

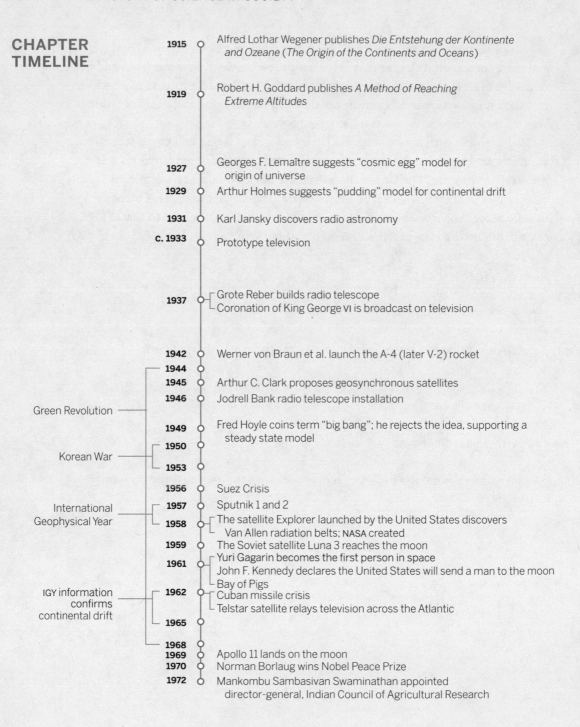

1915 Alfred Lothar Wegener publishes *Die Entstehung der Kontinente and Ozeane* (*The Origin of the Continents and Oceans*)

1919 Robert H. Goddard publishes *A Method of Reaching Extreme Altitudes*

1927 Georges F. Lemaître suggests "cosmic egg" model for origin of universe

1929 Arthur Holmes suggests "pudding" model for continental drift

1931 Karl Jansky discovers radio astronomy

c. 1933 Prototype television

1937 Grote Reber builds radio telescope
Coronation of King George VI is broadcast on television

1942 Werner von Braun et al. launch the A-4 (later V-2) rocket

1944

1945 Arthur C. Clark proposes geosynchronous satellites

1946 Jodrell Bank radio telescope installation

Green Revolution

1949 Fred Hoyle coins term "big bang"; he rejects the idea, supporting a steady state model

Korean War
1950

1953

1956 Suez Crisis

International Geophysical Year
1957 Sputnik 1 and 2

1958 The satellite Explorer launched by the United States discovers Van Allen radiation belts; NASA created

1959 The Soviet satellite Luna 3 reaches the moon

1961 Yuri Gagarin becomes the first person in space
John F. Kennedy declares the United States will send a man to the moon
Bay of Pigs

IGY information confirms continental drift
1962 Cuban missile crisis
Telstar satellite relays television across the Atlantic

1965

1968

1969 Apollo 11 lands on the moon

1970 Norman Borlaug wins Nobel Peace Prize

1972 Mankombu Sambasivan Swaminathan appointed director-general, Indian Council of Agricultural Research

1957: THE YEAR THE WORLD BECAME A PLANET

11

On October 4, 1957, Moscow radio announced that the Soviet Union had launched Sputnik and the distant stars were one step closer.

Dreams of space flight went back centuries, but only with the triumph of Big Science after World War II did the necessary components of scientific organization and technological development make it possible to launch objects into space. While individual achievement was still important, especially when Nobel Prizes were increasingly touted as a benchmark for the prowess of national science, the days of the lone scientist working in a homemade laboratory were essentially over. The bigger the question, the bigger the team created to solve it.

In addition, science was now a public commodity. Sputnik, which means "fellow traveler," orbited the Earth at an altitude of 900 kilometers and flashed across the sky at 29,000 kilometers per hour. It was a beacon in the night, proudly proclaiming the scientific and technological superiority of the Soviet Union. Publicity, prestige, and national pride were now linked to scientific achievement in a much more public way than seventeenth-century natural philosophers had found in their patrons' courts. Interest in science sent ripples through popular culture, with the appearance of movies and novels about spies stealing secret formulas or smuggling researchers out of foreign countries. Nevil Shute wrote about the end of the world by atomic holocaust in *On the Beach* (1957), and his book was turned into a film of the same title in 1959. The Nobel Prizes, which had for the most part been a minor news item before the war, now made newspaper headlines around the world.

The launch of Sputnik had not only political and military implications but also a psychological impact. For the first time, something created by human hands had left the Earth. While the concept of the Earth as a planet was firmly established, in practical terms it had been the limit of human experience until 1957. With the opening of the space era, it was possible to conceive of the solar system and the galaxy as something more than images seen through a telescope. What had been the realm of fantasy and science fiction became reality, expanding the potential zone of human activity enormously. It also hastened the spread of the idea of the Earth as a single biophysical unit – a single world, not a collection of continents or politically distinct regions that could function without regard for the state of other parts of the globe.

The same year that Sputnik was launched, 1957, was proclaimed the International Geophysical Year (IGY). The International Council of Scientific Unions (ICSU), an umbrella organization of various national scientific associations, initiated a massive research project to study both the Earth and the Earth's interaction with the universe. IGY ran from July 1957 to December 1958, with some of its projects running much longer. IGY and Sputnik also embody the ideological tension facing science and the scientific community in the modern world. The science of the IGY was to be cooperative, international, creative, nonpartisan, and open. Many scientists felt that these were the true characteristics of science, and they worked to get governments and the international community to celebrate this vision. The race to space, on the other hand, was characterized by secrecy, nationalism, and partisanship; moreover its military foundation was dedicated to destruction. Realizing that to compete militarily it was necessary to compete scientifically, military leaders around the world had little trouble finding scientists to work on their projects and generally enjoyed the support of the scientific community. Thus, the race to space, the arms race, and IGY represent different kinds of utility to the state. It is not surprising that IGY and Sputnik coexisted despite their apparently different characteristics, since both offered benefits to those who sponsored the work.

The significance of IGY and Sputnik must also be seen in the context of a world on the edge. The United States tested a non-deliverable fusion bomb in 1952. The Soviet Union tested its first partial fusion bomb, Joe 4, in 1953. In the same year, the Korean War ended in a stalemate. The Warsaw Pact, the Eastern Bloc equivalent of the North Atlantic Treaty Organization (NATO), was formed in 1955, heightening fears of another European war. And 1956 was particularly tense. The Egyptians, who were establishing ties with Eastern Bloc countries, planned to nationalize the Suez Canal, thus cutting off transportation to the East and oil supplies to Europe. To thwart this, Britain, France, and Israel planned a daring attack. Israel seized the Suez Canal and then surrendered it to the British and the French. When the American government failed to back them and the Soviet Union threatened intervention, the invading troops were forced to withdraw. Britain and France were left looking powerless and defeated. In the same year, a revolt against Soviet rule in Hungary was brutally suppressed, and the American military exploded the biggest hydrogen bomb ever at Bikini Atoll.

The launch of Sputnik 2 on November 3, 1957, confirmed the capabilities of the Soviet rocket program and clarified the military threat. Sputnik 2 was not only much larger than Sputnik 1 but on board was Laika, a Siberian dog. She lived in orbit for eight days until her oxygen supply ran out.

11.1 HYDROGEN BOMB TEST, BIKINI ATOLL, 1956

Weighing 508 kilograms, Sputnik 2 was large enough to be a weapon. Where World War II had expanded the zone of combat from the few kilometers of the trenches of World War I to the range of heavy bombers and V-2 rockets, the launch of the Sputnik satellites effectively eliminated the concept of the "front line" altogether.

And yet the launch of the first satellite fascinated the world's populations. Around the globe, people tuned into the radio beep that Sputnik broadcast and stood staring at the night sky hoping to see the satellite zoom by. Some people, particularly science fiction writers and fans, considered the launch of Sputnik the end of human childhood, a milestone that signaled the beginning of the next era of human development when we would leave the cradle of the Earth and move out to the stars.

The International Geophysical Year (IGY)

Lloyd V. Berkner (1905–67) and Sydney Chapman (1888–1970) initiated the idea of the IGY in 1950. Berkner was an electrical engineer by training but had wide interests in electronics, nuclear development, radar and radio, rocketry, and atmospheric science. In 1950 he was working on ionospheric physics in the Department of Terrestrial Magnetism of the Carnegie Institution of Washington; from 1951 to 1960 he was head of Associated Universities, Inc., which ran the Brookhaven Laboratories for the Atomic Energy Commission (AEC). Chapman, who had degrees in engineering, physics, and mathematics, had been interested for a long time in atmospheric science, having proposed a theory of ozone creation and depletion in 1930; in that same year, in association with Vincent Ferraro, he theorized that magnetic storms were caused when plasma ejected from the Sun enveloped the Earth.

Since Berkner had served with Admiral Byrd on his Antarctic expedition during the second International Polar Year in 1932–3 (the first occurred in 1882–3), he used the Polar Year as a model for this new and bigger project. The IGY was initiated in 1952 by the ICSU, an arm of the United Nations (UN) created to promote international scientific cooperation. To manage the actual operations ICSU established the Comité Spécial de l'Année Géophysique Internationale, with Chapman as its president and Berkner as vice-president. In the end, some 67 nations took part. In addition to gathering information, the ICSU established the World Data Centers, which archived and disseminated the information produced. The World Data Centers continue to function to this day.

The year 1957 was picked for IGY because it coincided with an expected peak in sun-spot activity, which followed an 11-year cycle. Other areas of research were aurora and air glow, cosmic rays, geomagnetism, glaciology, gravity, ionospheric physics, longitude and latitude determination, meteorology, oceanography, rocketry, and seismology. The result of this massive undertaking was one of the greatest records of the structure and function of planet Earth and its neighborhood.

The geological and oceanographic work provided the first global map of the structure of the planet. Although huge amounts of work had been done mapping the globe and geological surveys had been undertaken by various governments, especially the British, the material was not well integrated. Therefore, one of the tasks of the IGY was to coordinate masses of previously gathered data.

One of the major topics of interest that Berkner and the geologists hoped to address with this compilation of existing and new data was the question of ocean and continent formation. Although the specific approaches to the research question came primarily from nineteenth-century geology, particularly from Lyell and his *Principles of Geology*, new tools such as the seismograph were changing the way geologists and geophysicists studied the Earth.

Continental Drift

Like the shift from the Newtonian to the Einsteinian universe, the stability of geology was called into question when Alfred Lothar Wegener (1880–1930) introduced his ideas about continental drift in 1912. Trained as an astronomer and working mostly in meteorology, Wegener had noticed both the apparent "fit" of geological regions, such as South America with the west coast of Africa, and certain similarities in the distribution of animals and fossils; he combined these with paleoclimatological evidence. He was not the first to notice these relationships. Eduard Suess (1831–1914) had suggested the existence of supercontinents called Gondwanaland and Laurasia in his massive five-volume geological work *Das Antlitz der Erde* (*The Face of the Earth*), published between 1883 and 1909. Suess explained that the land bridges that connected the continents in prehistoric times had since disappeared, eroded, or collapsed into the oceans. In contrast, Wegener argued that the Earth had been molten in its earliest days and that it had contracted as it cooled, allowing lighter continental material to rise above the denser basalt of the crust, forming one supercontinent, which he called Pangaea. This continent subsequently broke up, and the parts representing the current continental masses "floated" apart. In 1915 Wegener published a detailed and expanded theory in *Die Entstehung der Kontinente und Ozeane* (*The Origin of the Continents and Oceans*). While the theory was popular (Wegener's book was translated into several languages and reprinted more than five times), it was almost completely rejected by the geological community. Although part of the rejection was territorial – what did a meteorologist know about geology? – there were good reasons to be skeptical. Wegener offered no reasonable mechanism for why the continents moved and seemed to suggest, among other things, that the various mountain ranges were about the same age if they were created during the formation of the supercontinent. It was clear to all geologists that the mountains of the world were not the same age.

In 1929 Arthur Holmes (1890–1965) attempted to revive Wegener's theory by postulating thermal convection in the Earth's mantle. Hot material was less dense than cool material, and thus it would rise. When it cooled, it would sink back again, like pudding in a pot. This cycle of heating and cooling might make the continents move. Holmes's idea received little serious attention, although continuing work in geology, revealing significantly more information about the structure of the planet, did not rule it out. The question of continent and ocean formation remained to be resolved.

During IGY, oceanographers worked hard to map the ocean floors. The mapping revealed not only a series of mid-ocean ridges but that near them the ocean floor was not uniform. It was newer near the mid-ocean ridges, with less sediment and younger rock. This ran against expectations. If the oceans were very old, then their floors should be thick with sediment laid down over the millennia. An even more surprising discovery was geomagnetic reversals. Using equipment developed to detect submarines, scientists mapped the magnetic orientation of the rock that made up the ocean floors. To their surprise, they found that the magnetic orientation of the mineral magnetite changed across the oceans. When rock was in a liquid state, the particles in it were randomly oriented toward the Earth's magnetic field, but once it started to solidify, the particles aligned along the magnetic lines of force, like millions of little compasses. Once the rock was solid, the orientation was frozen, pointing in one direction and unaffected by changes to the fields. The oceanic bands of rock made it clear that the fields had shifted and even changed places, with the magnetic north and south switching. These reversals were rapid and resulted in bands of rock with radically different magnetic orientation, which appeared to be laid out more or less symmetrically from the ocean trenches toward the continents. (See figure 11.2.)

With this new information, Holmes's idea received more attention, but "drifters" were still largely scientific pariahs in the 1950s, unable to publish papers in scientific journals and often risking career advancement. John Tuzo Wilson (1908–93), an influential geologist, IGY committee member, and president of the International Union of Geodesy and Geophysics (1957–60), did not initially support the theory of continental drift. However, by 1962 new information from IGY research published by Harry Hess (1906–69) and others, which gave scientific evidence for the spreading of the ocean floor, convinced him otherwise. Greater understanding of the ocean floor and the discoveries of features such as mid-oceanic ridges, geomagnetic anomalies parallel to the mid-oceanic ridges, and the association of island arcs with oceanic trenches near the continental margins, suggested convection might indeed be at work. Wilson's 1965 paper, "A New Class of Faults and Their Bearing on Continental Drift," helped transform the direction of geology. Although the term "continental drift" was eventually replaced by the term "plate tectonics" because it encompassed a broader geological system, in many ways Wegener's idea was vindicated.

11.2 MAGNETIC OCEAN FLOOR MAP

Another piece of evidence for continental drift came from space. Careful examination of the variation in the orbits of IGY satellites indicated gravitational anomalies, which suggested convection currents inside the Earth. This discovery was a by-product of another area of IGY research. Rockets and, more importantly, the instruments they carried were central to the study of the upper atmosphere and conditions in space. Many scientists wanted to use them as peaceful tools rather than weapons of mass destruction. James A. Van Allen (1914–2006), who was an alumnus of the Department of Terrestrial Magnetism at the Carnegie Institution and who had worked with captured German V-2 rockets at the end of World War II, was particularly keen on getting scientific satellites into orbit. Van Allen's work was the first American success to counter the Soviet Sputnik program.

The Green Revolution: Science and the Global Agriculture

One of the most significant applications of science to global problems in the Cold War era was the Green Revolution. The term "Green Revolution" was first used in 1968 by William Gaud (1907–77), the director of the United States Agency for International Development, although the scientific work had been going on since the 1940s. The term referred to the use of scientific principles and technology to increase agricultural production, particularly in the developing world. This was part of a belief that all the emerging problems of the modern world could be solved by science and technology. The leading figure in this revolution in agriculture was Norman Borlaug (1914–2009), a microbiologist. In 1944 Borlaug took a job working for the Rockefeller Foundation's International Maize and Wheat Improvement Center and went to Mexico to improve wheat production. At this time, Mexico was unable to grow enough wheat for its population, a major issue for a relatively poor nation.

The problem that Borlaug faced was simple to state but difficult to solve. To increase crop yield, wheat fields had to be fertilized with nitrogen fertilizers. The fertilized wheat produced more grain, but the weight of the kernels made the plant fall over, ruining the plant and reducing the yield. Borlaug's solution was to crossbreed a high-yield American wheat variety with a strong-stemmed dwarf wheat from Japan. By 1963, 95 per cent of Mexico's wheat consisted of Borlaug's hybrids and Mexico had become a net exporter of the grain.

Because of his success in Mexico, the Indian Ministry of Agriculture invited Borlaug to advise them on ways to increase their grain production. The Rockefeller Foundation and the Mexican government sent Borlaug and the wheat specialist Robert Glenn Anderson (1924–81) from the Canadian Department of Agriculture to meet with Indian officials and discuss ways to increase production. India and other countries in the region were experiencing low-level famine, but there was resistance to the new techniques. In 1965 Borlaug and Anderson arranged to send almost 500 tons of hybrid wheat seed to India and Pakistan, but the outbreak of war between the two countries slowed the project. To prevent starvation (and to encourage India to have closer ties to the United States), the United States sent 20 per cent of its wheat production to India in 1966. Ironically, the starvation caused by the war overcame local resistance to the new method of farming. Working with scientists such as Mankombu Sambasivan Swaminathan (1925–), who is considered in India to be the father of the Green Revolution, the basic hybrids were adapted for regional conditions. Swaminathan had been trained as a geneticist in India and was a post-doctoral fellow at the University of Wisconsin from 1952 to 1954. He worked on the Mexican project before returning to the Indian Agricultural Research Institute, where he was instrumental in developing the strains of wheat and rice for the South Asian region. By 1968 India, Pakistan, Turkey, Mexico, and several South American countries were using Borlaug's agricultural system. Wheat production nearly doubled and India became self-sufficient in grain production in 1974.

In 1970 Norman Borlaug won the Nobel Peace Prize for his work in this area. His efforts and the efforts of those who worked with him have saved the lives of millions of people. Borlaug was proud of his work, but he was also aware that he could be creating a kind of Malthusian trap if it was not properly used. In his Nobel acceptance speech he said:

> It is true that the tide of the battle against hunger has changed for the better during the past three years. But tides have a way of flowing and then ebbing again. We may be at high tide now, but ebb tide could soon set in if we become complacent and relax our efforts. For we are dealing with two opposing forces, the scientific power of food production and the biologic power of human reproduction.... Man also has acquired the means to reduce the rate of human reproduction effectively and humanely. He is using his powers for increasing the rate and amount of food production. But he is not yet using adequately his potential for decreasing the rate of human reproduction. The result is that the rate of population increase exceeds the rate of increase in food production in some areas.[1]

The Green Revolution was, like the IGY, an international effort by scientists to use science for the benefit of all humanity. Although certain aspects of agricultural change were used to garner support for the West as part of the Cold War, in the long run Borlaug and his supporters were more interested in people than politics. Modern critics have suggested that the Green Revolution did more for US agribusiness than for the adopting countries, and that monoculture farming can have bad environmental consequences including the loss of wilderness areas and a decline in biodiversity. On the other hand, this revolution allowed many countries in the developing world and especially in Asia to achieve food security and stability.

Mapping the Universe: The Steady State vs. the Big Bang

While the IGY program focused on the Earth and the Sun, some astronomers were creating a startling new image of the universe. By the time Sputnik was launched, much more was known about those stars that science fiction writers yearned to visit.

Astronomers had discovered that the universe was in motion. While this was not a new idea, no method had been developed earlier to observe the motion of

---

1. Norman Borlaug, "Acceptance Speech for the 1970 Nobel Peace Prize," The Nobel Prize, www
.nobelprize.org/prizes/peace/1970/borlaug/acceptance-speech/. © The Nobel Foundation, 1970.

objects outside the solar system, due to the difficulty of calculating the distance to stellar objects. In 1838 the astronomer Friedrich Wilhelm Bessel (1784–1846) was the first to measure the distance to a nearby star using stellar parallax. By measuring the tiny variation in the angle of observation from one side of Earth's orbit to the other (Bessel found 0.314 seconds of arc), he calculated that the star 61 Cygni was 10 light years from Earth. Although this method worked for close stars, it did not work for objects farther away since the light of distant objects had no measurable angle of intersection.

By the turn of the century, more than 500,000 stellar objects had been mapped. In 1917 George Ellery Hale (1868–1938) oversaw the installation of the 2.5-meter refracting telescope at the Mount Wilson Observatory. Named in Hale's honor, it remained the largest telescope ever made until the building of the 5-meter reflecting telescope at Mount Palomar in 1948. In 1926 the Bruce Proper Motion Survey began to trace the motion of stars by comparing photographic plates of the same region of space taken 25 years apart. The motion of the stars and the theory of relativity led Belgian astronomer Georges F. Lemaître (1894–1966) to argue in 1927 that the universe was created from an explosion of matter and energy, which he called the "cosmic egg."

Using data from the Wilson Observatory, in 1929 Edwin Powell Hubble (1889–1953) applied new methods based on the Doppler effect,[2] or the red shift of light, to calculate the distance to the Andromeda nebula as 930,000 light years away. He also found that most galaxies were moving away from the Earth and from each other. This fit Lemaître's idea, since it seemed possible to "rewind" cosmic history by tracing the movement of the nebulae backward and drawing everything together at some ancient time.

Fred Hoyle (1915–2001), one of the greatest astronomers of the era, objected to the cosmic egg theory, arguing along with Hermann Bondi (1919–2005) and Thomas Gold (1920–2004) that the universe was in a steady state. The Steady State model, presented in 1948, pictured the universe as uniform in existence, with no beginning or end, and looking much the same everywhere (on the large scale) and at every time. Hoyle, Bondi, and Gold dealt with the apparent expansion of the universe by arguing that matter was constantly created. Some

---

2. The Doppler effect, named for Christian Doppler (1805–53), explains the apparent rise and fall in the pitch of a moving sound. The pitch of the sound rises when it is moving toward you and falls when it is moving away from you. If an object, such as a star, is moving away from us, the wavelength of light it emits increases, making it look as if the light is shifted toward the red end of the spectrum.

considered the creation of matter from nothing to be an odd idea, but the amount required was very small (on the order of a few atoms per cubic light year per year) and it was no more philosophically unsettling than the question of the origin of the cosmic egg. Ironically, it was Hoyle who coined the term "Big Bang" as a derogatory label for the competing theory.

The debate between Steady State and Big Bang supporters was heated and often acrimonious. The evidence that tipped the argument in favor of the Big Bang (although there are still a few Steady Staters today) came from an unexpected source – radio waves from space. In 1931 Karl Jansky (1905–50) was researching radio interference for Bell Telephone Laboratories. He found three types of naturally occurring interference: local thunderstorms, distant thunderstorms, and an unknown but constant source, which he initially thought might come from the Sun. This was a reasonable idea, since back in 1894 Sir Oliver Lodge (1851–1944) had claimed that the Sun emitted radio waves, although he had been unable to detect them. However, after a year of study Jansky determined that the unknown source was actually the Milky Way. Our home galaxy was emitting radio waves! When Jansky finished his investigation, Bell assigned him other research, and he never followed up on his discovery. He died of a heart attack at the age of 44, but in his memory the unit of energy flux, or "radio brightness," is called a Jansky (Jy).

Jansky's discovery was largely ignored, but Grote Reber (1911–2002), a radio engineer, was intrigued with the idea of radio waves from space and built a 9.4-meter parabolic dish antenna in 1937 in order to capture these emissions. He set out to create a radio map of the heavens. Between Jansky's serendipitous discovery and Reber's work, the field of radio astronomy was born. Because of the interest in looking at extreme events through physics (the interior of stars, for example) and the relationship of general relativity to the structure of the universe, money for radio astronomy became available after World War II, as did stocks of leftover electronic equipment, often available as cheap military surplus. In 1946 the first major radio telescope installation – a 66 meter parabolic aerial – was built at Jodrell Bank in Britain. (See figure 11.3.) Radio astronomy took a leap forward when Martin Ryle (1918–84) introduced the first radio interferometer. Essentially a system of two receivers hooked together, it provided a much sharper image and more detailed information about radio sources. With simple interferometers and arrays of many radio telescopes linked together, radio astronomy became as good at resolving light sources as optical telescopes. Reber, Ryle, and others began mapping the heavens, creating a much more complex picture as things visible only by radio detection were added to centuries of optical observations.

11.3 JODRELL BANK TELESCOPE

Source: Andrew Barker / Shutterstock.com.

The Space Race

The technical advances that made radio astronomy possible also made access to space, especially in local Earth orbit, increasingly attractive. The race to space, however, was more than an intellectual footrace. It was also a battle between scientific styles. It is easy to confuse the search for the most productive style of research with the political battle between the Western democracies and the totalitarian regime of the Soviet Union, but the relationship between scientific research and the demands of the state, regardless of the structure of government, remains complex. Scientific success does not depend on a free populace, but the translation of discovery into products for the larger population has usually been slower under totalitarian governments. Yet, under both democratic and totalitarian governments, research deemed to be in the national interest is developed and controlled by bureaucracies that may not even understand the science involved, that measure success or failure by objectives met rather than discoveries made, and that ensure that money rules the course of research.

The story of rocketry that leads to Sputnik is one of despots and dreamers, technocrats, technicians, politicians, and researchers. While the race to space did not create the integrated system of scientific research, it is the greatest example of the power of such integration. The development of nuclear weapons was in many ways a more complex integration of research and the demands for a "useful" final product, but it was shrouded in secrecy and was presented as so advanced and esoteric that it was accessible only to geniuses. The rocket race was, in contrast, a very public demonstration of scientific prowess that led to a revision of the place of science, and especially scientific education, in the industrialized world.

The first rockets appeared around 1150 when the Chinese used gunpowder to propel them. They were later developed into a weapon at the siege of Beijing in 1232 and made their first appearance in Europe in 1380 at the battle of Chioggia between the Genoese and the Venetians. Although small rockets continued to be made for fireworks and military purposes, they were largely overshadowed by cannons and artillery. The exception to the general trend was in Russia.

In 1881 Nikolai Kibalchich (c. 1853–81) built the bomb that killed Tsar Alexander II. He made it in the chemical laboratory of the St. Petersburg Technological Institute, founded by Alexander's father to help propel Russia into the scientific and industrial age. After his arrest, Kibalchich spent the time before his execution designing rockets, including a design for a rocket plane and a passenger rocket. Kibalchich was not an aberration; Russian interest in rocketry was long-standing. One hundred years earlier the *Raketnoe Zavedenie* (Rocket Enterprise) was created to design and manufacture rockets. The first missile unit in the Russian army was formed in 1827, and in 1867 the engineer Konstantin Konstantinov opened a rocket-manufacturing plant in St. Petersburg.

The same year that Alexander II was assassinated, Konstantin Tsiolkovsky (1857–1935), a self-taught physicist, sent a paper to the Russian Society for Physics and Chemistry in St. Petersburg that dealt with the kinetic theory of gases and the mechanics of living organisms. He was told that his ideas were already well

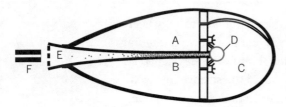

A. Liquid Oxygen Tank   D. Combustion Chamber
B. Liquid Hydrogen Tank   E. Exhaust Nozzle
C. Cargo or Crew Area   F. Control Surfaces (guidance)

11.4 LIQUID FUEL ROCKET

Illustration based on Konstantin Tsiolkovsky's 1903 sketch of a liquid fuel rocket.

known to scientists, but rather than being discouraged he continued to pursue his interests. His driving passion was the conquest of gravity. Throughout the 1880s he contemplated what life would be like in zero gravity, and in 1903 he published the mathematical treatise "Exploration of Cosmic Space with Reactive Devices" on orbital mechanics and the principles of rocket propulsion. He also sketched a liquid propellant rocket, which used liquid oxygen and liquid hydrogen as fuel. (See figure 11.4.)

Tsiolkovsky was not the only person working on advanced ideas in rocketry in this period. Robert H. Goddard (1882–1945) in the United States began serious experimental work in 1909 when he was a student at Clark University. He eventually published his PhD thesis, *A Method of Reaching Extreme Altitudes*, with financial help from the Smithsonian Institution. Goddard was a persistent and ingenious inventor, but he was also secretive and isolated. In 1926 he moved to a ranch near Roswell, New Mexico, to get away from reporters and other rocket enthusiasts and carried out foundational work on liquid-fuel rockets. While he was granted more than 200 patents, his work was little known at the time.

Rocket Science in Germany and the Soviet Union

Through the 1920s and 1930s the general public became interested in rocketry. In 1930 the legendary science fiction editors Hugo Gernsback (1884–1967) and David Lasser (1902–86) founded the American Interplanetary Society (AIS). The AIS corresponded with Robert Esnault-Pelterie (1881–1957), the leading rocket engineer in France, and with members of the German amateur rocket society Verein für Raumschiffahrt (VFR), as well as financing liquid-fuel rocket work at home. Despite this popular interest, however, most institutions and governments paid little attention to rockets. To a certain extent, this lack of interest was due to the rise of powered aviation, which was more immediate and personal since it offered human flight, but it was also because dreams of flight into space seemed both preposterous and pointless. Two governments proved the exception to this disinterest: the German government under Hitler and the Soviet government under Stalin.

The German story is better known but was ultimately less successful. In the 1920s the reorganized German army developed strategies for the next war. They chose a mechanized, highly mobile approach that used aircraft as a form of long-range artillery. This left a tactical gap between heavy artillery, with a maximum range of about 100 kilometers but an effective range much less than that, and bombers that could fly hundreds of kilometers to attack targets but were expensive and required a great deal of logistical control. The military planners began to consider rockets, which were not prohibited under the terms of the Treaty of Versailles, and the Ordinance Ballistic Section of the German army was given the task of developing a liquid-fuel rocket to fill the gap between artillery and air power. The actual job fell to Walter Dornberger (1895–1950), who had studied ballistics at the School of Technology at Charlottenburg.

In 1929 Dornberger visited the "rocketport" used by the rocketeers of the VFR and met Wernher von Braun (1912–77). He encouraged von Braun to complete his bachelor of science degree in engineering at the Berlin Institute of Technology. While earning this degree, which he finished in 1932, von Braun worked with Hermann Oberth (1894–1989), the leading rocket scientist of the day, on the construction of liquid-fuel rocket engines. Oberth, who had written on the physics of space flight in the 1920s, was von Braun's original inspiration. In 1934 von Braun completed a degree in physics at the University of Berlin.

By this time Dornberger was head of Research Station West at Kummersdorf, but German interest in rockets was waning as other aspects of military technology, backed by an industry that already existed to produce the materials, took center stage. With von Braun, Dornberger established a test facility at Peenemünde in 1937; it opened in 1939. Their efforts were limited by budget constraints, but in 1942 they launched the first A-4 rocket.

As the war began to go against Germany, Hitler looked for weapons that would provide a technological fix to the problem of military failure. The V-1 rocket, which consisted of a jet engine on a winged bomb, was so slow it could be shot down by fighters, but the A-4 as a ballistic weapon was unstoppable. With a maximum range of about 400 kilometers and a payload of 1,000 kilograms, it was the kind of superweapon that Hitler wanted. Put into mass production as the V-2, or "Vengeance Weapon," its destructive power was significant. Although by 1945 some 300 V-2 rockets were being assembled each month by slave labor at the Mittelwerk factory near Nordhausen, they had virtually no strategic or tactical effect on the outcome of the war. The rocket designers were pleased with the success of the A-4, but they were planning much greater things. The A-9 was to be a two-stage rocket capable of intercontinental flight, the A-11 a three-stage rocket capable of lifting a pilot into space.

As the German military collapsed under attack from east and west, von Braun and the "Rocket Team" contemplated the end of the war. If they stayed at Peenemünde and Nordhausen, they would be in the Russian zone of control, so they decided to surrender to the American army. In February 1945 von Braun and more than 500 people from the rocket program, plus mountains of documents, traveled south across the devastated country. At the beginning of May they contacted and surrendered to the American military in Bavaria. When the American command realized what kind of prize they had in hand, they staged a special mission to clean out everything of use from the Mittelwerk factory, sending the equipment, unused V-2 rockets, and eventually many of the German scientists and engineers to the White Sands Proving Grounds in New Mexico.

The Americans scooped most of the rocket program out from under the noses of the advancing Russian forces. Only one major German scientist, Helmut Gröttrup (1916–81), and a small group of workers went to the Russians, but Gröttrup's work did not contribute directly to the next stage of rocket development. When Russian experts arrived at Nordhausen and Peenemünde, they concluded that German manufacturing efforts were more advanced but not more theoretically sophisticated than the Russian. The loss of the German rocket specialists did not appear to pose a risk until the first atomic bomb exploded over Hiroshima on August 6, 1945. The implications of a sudden boost to rocket development by the United States changed the whole complexion of the technological race.

Rocket development in the Soviet Union brings to the fore the whole issue of the relationship between science and the state. Just as tsars of an earlier era had hoped to catapult Russia into the industrial age by luring selected industrialists and founding scientific societies, the Soviet regime hoped to use rocketry and space flight to help propel Soviet society into a leading role in science and technology. There was a particular match between rocketry and the ideology of the revolutionaries who toppled the Russian monarchy. Alexander Bogdanov's (1873–1928) popular novel *Red Star* (1908), reprinted in 1917, linked communist ideology and space flight with a workers' paradise created by Martians. Lenin himself had argued for the promotion of science and technology and political tolerance for great scientists because he recognized one of the lessons of World War I: the side with the best technology wins.

The Soviet government created new research institutions, and while many were short-lived, interest in rocketry remained strong. In 1924 the Central Bureau for the Study of the Problems of Rockets (TSBIRP) was created to coordinate research and focus attention on the military development of rockets. A private group also formed that year, the All-Union Society for the Study of Interplanetary Communications (OIMS); it was a parallel to the AIS in America and the VFR in Germany. In 1927 TSBIRP and OIMS hosted the Soviet International Exhibition of Rocket Technology in Moscow.

When Stalin came to power, he was far less tolerant of independent research and purged many scientists and engineers, sending them to jail, exile in the gulags, or death. At the same time he was prepared to support science and technology in certain areas. The Academy of Sciences was expanded, and its budget rose from 3 million rubles in 1927 to 175 million in 1940. Enrollment in technical schools also rose dramatically. Stalin set up a system of *sharashkas* (prison design bureaus) to utilize the talent of the jailed scientists and engineers. One such scientist was Sergei P. Korolev (1906–66), who graduated from the Kiev Polytechnic Institute but was jailed and sent to work in the Kolmya gold mines, one of the worst of the gulags. Korolev was saved by the great aviation designer Sergei Tupolev, himself a prisoner, who arranged for him to be transferred to TSKB-39 *sharashka*, which was filled with aviation specialists who continued their work, even under appalling conditions. The war with Germany put an abrupt halt to most of this work; however, one of the prison camp inmates, Georgy Langemak (1889–1938), created a militarily useful rocket, the Katyusha, which was fired by multiple launchers carried on trucks.

After the war Soviet efforts to turn the dreams of rocketry into reality began to bear fruit, but it was not until after Stalin's death in 1953 that work began to accelerate. In 1954 A.N. Nesmeianov (1899–1969), president of the Soviet Academy of Sciences, declared that it was feasible to send a rocket to the Moon or to place satellites in orbit. The Academy created a high-level Commission for Interplanetary Communications, which signaled significant support for the project of satellite launches and for the scientific foundation for intercontinental ballistic missiles (ICBMS). Korolev, now working on long-range missiles under better conditions for the government, was directed to turn his attention to ICBMS in 1953, the same year he became a Corresponding Member of the Soviet Academy of Sciences. By 1955 there was a new test facility at Tyuratam, and test firing of rockets was almost routine. The hopes of the project lay with Korolev's R-7, a stubby but massive rocket that used kerosene and liquid oxygen as the propellant. It combined 20 rocket engines in clusters and produced more than a half-million kilograms of thrust.

The first launch failed, but a second rocket launched on August 3, 1957, traveled from the launch site to land in the Pacific Ocean off the Kamchatka Peninsula. On October 4, Sputnik was put into orbit. Premier Khrushchev congratulated the team and in the following days began a propaganda campaign based on their success.

The United States Enters the Space Race

The launch of the first artificial satellite was not a complete surprise to the West; in fact, Moscow had announced the ICBM launch and even made public on October 1 the frequency on which Sputnik would broadcast. Yet many in the West, particularly in the American rocket programs, were startled. Both the Soviet success and the fragmented state of American efforts highlighted a basic problem with American research.

During the war, the OSRD had coordinated civilian research efforts and been the conduit for federal funding, which ended with the war. The OSRD, which had overseen significant work on a range of wartime projects, was in danger of being closed down.

Vannevar Bush, one of the heads of the ORSD instrumental in persuading President Roosevelt to mobilize American civilian scientists for war work in the first place, wanted to continue the organization in some form after the war. Although a peacetime version of the OSRD seemed desirable, it was considered to be unconstitutional since it required federal funding not controlled by elected representatives. For this reason, a number of bills proposing the establishment of a research organization failed.

A recommendation from the military led to the creation of the Research Board for National Security. It was not well received. Civilian scientists feared it was too open to interference from the military and contractors, while many politicians and bureaucrats felt that it would be too powerful an organization to operate without the oversight of the president or Congress. It was terminated in 1946. While virtually all the parties involved – the executive branch, Congress, the scientists and their universities, and the military – were interested in continued research, they could not decide on a postwar format. Scientists wanted federal money with few or no strings attached, while the government by law controlled funding and by preference wanted a high level of oversight. This left researchers with no national organization, and large-scale research funding fell to the military. By 1950 more than 60 per cent of all federal money for research came from the military.

The American rocket program – or more accurately, programs – grew slowly during this period of muddled research policy. When Congress restructured the military in 1947, creating a single Department of Defense, major missile research was divided between the army and the air force. The army, for whom von Braun was working, had control of tactical missiles, while the air force worked on strategic missile programs. But the air force, being composed of pilots, was naturally more interested in aircraft than missiles and concentrated on the development of long-range bombers and fighter planes.

When Dwight Eisenhower was elected president in 1952, he wanted to combat the rising cost of the federal budget and keep the United States out of "brush fire" wars, such as the struggle in Vietnam over French colonial rule. More than 57 per cent of the federal budget was going to the military, and military demands for bigger and more powerful weapons continued to climb. It was hard to limit military spending during the Cold War. Because the United States could not or would not match the size of Soviet ground forces, its general policy was to use air power and technical superiority to counter Soviet numbers. This meant more than bombers with nuclear weapons; it meant missiles. As intermediate range ballistic missiles (IRBMs) and ICBMs increasingly became the objective of research and development, a competition between the different branches of the military gave rise to a series of different missiles, each with different capabilities and championed by competing organizations. The army's Redstone (a direct descendant of the V-2) and Jupiter rockets as well as the air force's Atlas, Titan, and Thor rockets were all rushed into development. Although they were used in the race to space, these rockets were actually designed to replace weapons such as the Minuteman and the Polaris missiles.

The "civilian" space program, Project Vanguard (actually run by the Naval Research Laboratory), was the public face of the American race to space. It did not fare well, as many scientists and engineers were absorbed by other projects, particularly in nuclear research and ballistic missile development. Regarded as a second-class project by the military and a threat to funding, it was also plagued by technical problems. When its funding requirements rose from $20 million to $63 million in 1956, it appeared to be a project out of control.

CONNECTIONS

NASA as Big
Science

In 1957 the Soviet Union beat the United States into space when the Soviets launched Sputnik 1 and Sputnik 2. This challenged the widely held belief that American technology, and technology in the West more generally, was well ahead of that being used in the Soviet Union. Western technology was in fact ahead of Soviet technology, but the democratic and somewhat chaotic nature of US research meant that it could not be as focused as the state-directed Soviet system. In the United States, various parts of rocketry were being developed by many different groups, including the air force, the navy, and the army, as well as the National Advisory Committee for Aeronautics and a number of different universities. In many ways, they had forgotten the lessons of the Chemical Warfare Service and the Manhattan Project, allowing a sort of free-for-all when it came to space exploration. With the launch of Sputnik, this became a matter of national pride and security, so President Eisenhower, taking the advice of the National Science Foundation and the National Academy of Sciences, created the National Aeronautics and Space Administration (NASA) in 1958 to bring all of the major rocket programs together under a civilian administration.

With one umbrella organization and significant funding from the state ($20 billion, equal to about $210 billion today), the Apollo project was established to land an American on the Moon. This was Big Science as heroic adventure and certainly placed American space exploration ahead of the Soviets. NASA went on to orbit satellites including the

Despite all these problems, when the R-7 sent Sputnik 1 into orbit, the United States was not significantly behind the Soviet Union in development and was ahead in guidance, miniaturization, and weapon building. A crash program by the JPL and the Army Ballistic Missile Agency built Explorer 1 in 84 days and launched it on January 31, 1958. It was followed by Vanguard 1 on March 17, 1958. The military missile programs were successful, although the division of labor limited their achievements to certain specific goals. None of that mattered,

Hubble telescope, create Skylab, send probes to other planets, and participate in several other space exploration programs.

If the purpose of NASA was only to collect Moon rocks and place a flag on the Moon, public concerns about its large budget would be justified. However, NASA has played an incredibly important role in the development of a high-tech economy through science that is only possible in these large, well-funded teams. The efforts of NASA required the work of many people, but the organization did not treat everyone equally. Women and people of color who worked behind the scenes doing everything from working out the complex calculations of orbital mechanics to making space suits, rarely received any acknowledgment of their contribution to the space race. In addition to the space-based tech-nologies – ranging from GPS satellites, aviation navigation, and radar systems, which were a direct development of the race to space – NASA has also developed or co-developed a huge range of products that have become consumer goods. These include infrared ear thermometers, advanced radial tires, memory foam, solar cells, and freeze-dried food. The list of technologies for aircraft alone would be over 50 pages long. Equally important to the modern techno-science world, because NASA is publicly funded, it has a mandate to make public its findings. In an age where science is becoming more and more proprietary, NASA stands out as the best of public Big Science. In 2014 alone, NASA released more than 1,500 pieces of software (including project planning tools and sound modeling) to the public for free.

NASA was created as an organization that fosters invention and inquiry. NASA is the largest collection of scientists and engineers in one organization in human history. In 2015 more than 10,000 people worked in the science and engineering sectors of NASA, forming a vast pool of talent. It trains people to take on enormous projects and those skills percolate out to the rest of society. While some people might say that the value of knowing more about distant stars and planets is low, few people would say that the world would be better off if there were 10,000 fewer scientists and engineers. Far more than the Royal Society or the APS, NASA has produced useful knowledge. It has lived up to its vision statement: *To reach for new heights and reveal the unknown so that what we do and learn will benefit all humankind.*

however. The launch of Sputnik meant that American efforts looked second-best, the line taken by many leading media outlets such as *Life* magazine. A national outcry followed, and blame for the perceived failure of American science was cast far and wide. President Eisenhower was forced to confront a political dilemma. Should the United States collectivize its scientific effort as the Soviet Union had done and commit even more federal resources to military spending in a visible effort to surpass the Soviets? How had the Soviets managed to

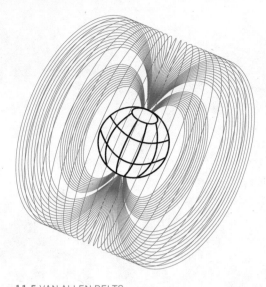

11.5 VAN ALLEN BELTS

The Van Allen belts form a protective electromagnetic field around the Earth.

achieve so much in such a short time? Was there a problem with American education?

There was much soul-searching about this problem. While science and engineering training had vastly increased since World War I, many people, such as the editors of *Life* and Vannevar Bush, wanted to see a significant increase in spending on education. The problem was politically sensitive because education was a state responsibility and therefore outside the purview of the federal government. Eisenhower was reluctant both to commit large sums of money and to be seen to trample on state rights. In 1958 Congress enacted the National Defense Education Act as a compromise between those who wanted no federal spending on education and those who wanted a massive increase. The bill authorized an expenditure of about $1 billion over four years. It created a $295 million loan fund for students in financial need and allocated $280 million of matching federal grants aimed at purchasing equipment for science, mathematics, and language training. There was an additional $60 million fund for 5,500 graduate fellowships in areas related to national defense.

When James Van Allen and his team successfully launched Explorer on January 31, 1958, they demonstrated that American efforts were not so far behind the Soviet's. Van Allen's satellite project began back in 1955 as part of the IGY research on radiation. Although the American satellite was far smaller than either Sputnik 1 or Sputnik 2, weighing only 14 kilograms, it nonetheless established the American capability to orbit satellites. As a tool for scientific research rather than for national security, its instrument package contained a radiation detector that demonstrated the existence of bands of radiation above the Earth. Named after their discoverer, the Van Allen radiation belts range from 650 kilometers to 65,000 kilometers and circulate along the Earth's magnetic field. (See figure 11.5.) Explorer was both a scientific success and a political statement, but it lacked the status of being first.

Despite Explorer getting into space and the successful launch in 1958 of the Atlas missile, many Americans were concerned about the technological gap between the United States and the Soviet Union in the development of ICBMs.

Missiles had made long-range bombers obsolete, so the air superiority the United States had enjoyed now disappeared. The issue was further inflamed when journalist Joseph Alsop (1910–89) published articles claiming that by 1963 the Soviet Union would have 2,000 operational ICBMS compared to only 130 planned by the American military. The American Central Intelligence Agency (CIA), for various reasons, also predicted a large "missile gap" by 1963, despite little evidence. (In actuality, by 1961 the Soviet Union had deployed only 35 ICBMS.) Eisenhower was forced to increase spending, including $186 million for a spy satellite program being developed by the Advanced Research Projects Agency (ARPA) under the direction of the CIA.

The Cold War Heats Up: Communication Satellites and Television

While Eisenhower was confident that American military efforts would eventually overtake Soviet production, secret missile programs did little to quell public concern. What the United States needed was a major project to surpass the Soviet effort. Such a project had to fulfill three requirements: it had to put the United States in the lead in space, it had to establish the superiority of American science and scientific organization, and it had to be something that could excite the public. The race was not just a technological race but a fight for international prestige.

The first step was Eisenhower's creation of NASA on October 1, 1958. It brought together the National Advisory Committee for Aeronautics (with 8,000 employees and a budget of $100 million) with the Langley Aeronautical Laboratory, the Ames Aeronautical Laboratory, and the Lewis Flight Propulsion Laboratory. It went on to include the space science group at the Naval Research Laboratory (home of Project Vanguard), the JPL run by Caltech for the army, and the Army Ballistic Missile Agency, the center where Wernher von Braun and his team were located. The first director was T. Keith Glennan (1905–95), who was trained as an electrical engineer. He had worked at the Columbia University Division of War Research and later at the American Navy Underwater Sound Laboratory during the war. When he was chosen to be director, he took a leave from his job as president of Case Institute of Technology. During his term, he consolidated almost all nonmilitary space research and development under the NASA umbrella.

The national outcry over Sputnik had led to the creation of a strange hybrid of military and nonmilitary activity in NASA. One of its reasons for existing was to do research of military significance but not under the control of any branch of the existing military. To go along with it, Congress created two new standing committees, the Senate Committee on Aeronautical and Space Sciences and the House Committee on Science and Aeronautics. These committees wrestled with the complex problems of civilian-military relations and objectives, the degree and method of cooperation with space programs in other countries, and, most of all, the money.

In 1959 the Soviet Union upped the ante once more with the announcement of its Lunik series of experiments. Although some early attempts to send a rocket to the Moon had failed, Korolev and his team finally succeeded in producing a rocket with enough thrust to escape the Earth's gravity. Luna 1 was aimed at the Moon but missed the target and eventually went into orbit around the Sun. In good propaganda style, the Soviet press hailed the launch as a complete success as the first artificial planet and renamed it *Mechta* (Dream). Luna 2 reached the Moon on September 14, while Luna 3 passed behind it and took the first photograph of the far side, which, because its rotation matches its orbit around the Earth, had never been seen.

The American response was von Braun's Pioneer 4, which was launched July 16, 1959. The effort was at best a partial success. The rocket did lift off and gained sufficient velocity to escape the Earth, but guidance problems caused it to go off course and miss the Moon by 60,000 kilometers! While this was not a huge error in astronomical terms, it was an embarrassment for von Braun and his team.

In 1960 John F. Kennedy defeated Richard Nixon to become president. At 43, he was the youngest president ever elected and the first to exploit the power of television for electoral purposes. On May 25, 1961, in a speech to Congress, he declared that it was "time for a great new American enterprise – time for this nation to take a clearly leading role in space achievement, which in many ways may hold the key to our future on earth."[3] That enterprise was to send a man to the Moon and bring him home safely. It would not be cheap. An early estimate for the cost of the Moon project was $7 billion. When the Apollo program ended in 1973, it had cost almost $20 billion out of a total of $56.6 billion spent on NASA.

....

3. John F. Kennedy, "Special Message to Congress on Urgent National Needs," May 25, 1961, John F. Kennedy Presidential Library and Museum, audio, 46:01, www.jfklibrary.org/asset-viewer/archives /JFKWHA/1961/JFKWHA-032/JFKWHA-032.

President Kennedy needed mass support for such a major undertaking, and, indeed, much of the point of the race to the Moon was the mass exposure and approval it would garner. When he made his 1961 declaration, Kennedy was speaking not only to Congress but to the whole nation. His words were recorded and broadcast across the country. One of the significant differences about national and international politics in the second half of the twentieth century as compared to earlier eras was that they were propagated and transmitted to a huge audience. Radio broadcasts had become a key component of government activity through Hitler's rally speeches, Churchill's wartime messages, and Kennedy's declaration of American will. Radio communication had undergone significant improvement through the 1940s and early 1950s, but the problem of long-distance transmission needed to be solved before Kennedy's mass democracy became a possibility. Communication satellites, made possible by the developing science of rocketry, offered a tantalizing solution to this problem. As early as 1945 Arthur C. Clarke (1917–2008) proposed the idea of geosynchronous communications satellites. His paper, "Extra Terrestrial Relays" in *Wireless World*, argued that a satellite placed at about 35,900 kilometers above the surface of the Earth could have an orbit that kept it stationary over a point on the surface, making it a very useful communications link.

Proposals for military reconnaissance satellites were made as early as 1950, but the cost and problem of the international legality of satellites traveling over foreign countries raised serious concerns. As the demand for information rose steadily during the Cold War, the U-2 spy plane was developed and began over-flights of Soviet territory in June 1956. In 1957 a U-2 returned with photographs of the Tyuratam launch facility. The theory of satellites for communication and observation had suddenly become a reality.

While the words of world leaders and soap manufacturers were being broadcast by radio, many major events of the era were televised, often transmitted live from around the globe. In tandem with the Cold War and the race for space, television came of age in the postwar era, one of the great examples of the transformation of a scientific object into a commonplace device. Each television is a microcosm of modern science and represents one of the unexpected products that research has given us. Like the earlier invention of printing, then of the intersection of telegraphy and the newspaper, television has been used to introduce many of the products of modern science.

Television does not have a single inventor but rather hundreds of contributors. The idea of transmitting images electrically developed along with the telegraph, and working methods were devised as early as 1875. The first practical image transmitter was demonstrated by Paul Nipkow (1860–1940) in 1905, based on an idea he had in 1884. It involved a spinning disk with a spiral of holes. Light passing through the holes fell on an image and was converted to an electric signal that controlled the intensity of a lamp set behind a matching rotating disk at the receiving end. In effect, the first perforated disk scanned the image and the second synchronized the light impulses and turned them back into the image. Using a disk system, the first still picture was broadcast by radio in 1907. By 1924 a moving image could be transmitted, but the image size was tiny, about 2.5 centimeters.

The invention of the Audion amplifier in 1906 by Lee Deforest (1873–1961) opened the door to a number of electronic devices. The Audion was used to boost an electrical signal in a cathode ray tube or vacuum tube. Numerous developments in tube technology eventually allowed Philo Taylor Farnsworth (1906–71) to develop an electronic picture tube in 1927; in that same year Bell Telephone Laboratories broadcast pictures between Washington, DC, and New York. In 1923 Vladimir Zworykin (1889–1982) significantly improved camera technology when he invented the "Iconoscope," a camera tube that combined lenses and used a photoelectric mosaic to capture the image. The prototype was demonstrated in 1929 and was manufactured by RCA in 1933. By 1935 television was being broadcast in Britain, Germany, and France; in 1936 the Olympics in Berlin were televised. The coronation of King George VI was broadcast in 1937. In 1939 Franklin D. Roosevelt became the first president to make a televised speech. Although broadcasts were very local, more than 20,000 television sets were sold in Britain by the end of 1939.

The demand for television was curtailed by the war, as both the people and materials needed were diverted to war work. With the end of the war and with the creation of uniform technical standards, there was a huge demand for television, which was spurred on even further with the introduction of color broadcasting starting around 1954 in the United States and 1967 in Europe – just in time to broadcast the events of NASA's Apollo program. Almost every move of the NASA astronauts was carried out in front of television cameras. Even if the public did not always understand what they were being shown, television changed the place of scientists in society and certainly attracted young people to the subject.

The ability through the space program to put large objects in orbit made communication satellites a reality. In 1960 the United States launched Echo, the first passive communications satellite, which was little more than a silver balloon. This was followed in 1962 by Telstar, which relayed the first transatlantic television broadcast.

After the launch of Telstar, equipment that had been used with Echo was no longer needed for commercial work. Arno Penzias (1933–) and Robert Wilson (1936–) were allowed by Bell to use a microwave detector (called a "horn antenna" because it looked like a ram's horn) for radio astronomy. Working on emissions in the microwave region, they found a constant hiss, somewhat like an FM radio tuned to an empty frequency. While this had not interfered with communications, it was a problem for astronomical work. After eliminating potential local sources of interference (such as pigeons in the receiver), the noise remained. It wasn't in the equipment. It wasn't radiation from nuclear test fall-out or even from the Sun. They could detect a faint bit of radiation even in empty portions of the sky.

Both frustrated and intrigued, Penzias and Wilson looked for a theoretical explanation for the persistent background noise. They contacted Robert Dicke (1916–97), who was working at Princeton University on theories about the Big Bang. He suggested that a kind of residual "noise" in the form of low-level background radiation would be left over from the original explosion. Penzias and Wilson had the evidence in hand. The universe was producing emissions like that of a black body (as investigated by Kirchhoff and Planck; see Chapter 8) at a temperature of around 3° Kelvin. The cosmic background radiation was uniform in all directions, meaning that the universe was expanding uniformly (fitting with Hubble's and Lemaître's ideas) and further that the universe had started at the same initial temperature. Although alternative reasons for this were presented by the Steady State supporters, the Big Bang eventually became the standard model for cosmology, in part because of this evidence. Penzias and Wilson went on to win the Nobel Prize in Physics in 1978, sharing the prize with Pjotr Leonidovich Kapitsa (1894–1984).

While physicists and astronomers concentrated on the big picture, NASA was still working on getting around the neighborhood. The first practical question that had to be answered about a trip to the Moon was whether a human could live in space. Some technical aspects were well understood, such as the lack of atmosphere and the general physiological impact of launches, but others, such as the effects of weightlessness and exposure to radiation, were not clear. Answers to these questions required manned flights. Both the Soviet and American space programs experimented with animals, and then moved rapidly to be first to put a man in space. The Soviets won. The first man in space was Yuri Gagarin

(1934–68), who orbited the Earth on April 12, 1961. NASA sent Alan Shepard (1923–98) on a suborbital flight on May 5 of the same year.

Although space travel now seemed a possibility, the continued existence of humanity was thrown into doubt. In 1961 an American-backed invasion of Cuba failed. The Bay of Pigs fiasco was a major embarrassment and increased tensions between the United States and the Soviet Union. A year later the world held its breath through the Cuban Missile Crisis. American spy planes recorded the construction of missile sites in Cuba that could be used to launch nuclear weapons. Since Cuba was under 150 kilometers from the American coast, there was no need for ICBMs to deliver the weapons. President Kennedy ordered the blockading of Cuba by the navy, as Nikita Khrushchev proceeded with a shipment of military material to Cuba. A tense standoff followed. The Crisis, which we now know was characterized by a lack of information, military mistakes, and confusion, brought the world to the brink of war as both sides put their forces on high alert. Fortunately, secret negotiations ended the standoff. The Soviet Union agreed to remove its weapons from Cuba, and the United States quietly took its missiles out of Turkey. Just as the world started to recover, in 1963 President Kennedy was assassinated.

After so many problems, the United States was ready for some good news. Kennedy's commitment to get to the Moon was endorsed by the following administration and became one of his legacies. The Kennedy Space Center was named in his honor. NASA and the Apollo program became the largest nonmilitary project ever undertaken by the United States. At its height, it employed almost 30,000 people directly and thousands more through contractors and represented about 5 per cent of the federal budget.

One of the main accomplishments of Apollo was the production of the Saturn rocket series. These liquid-fueled rockets, particularly the Saturn V, were the workhorses of the space program. The program was set back in 1967 when a fire on Apollo 1 killed the crew on the launch pad, but in 1968 Apollo 8 orbited the Moon, essentially resolving all the technical issues about getting there and back, short of landing. Apollo 11 touched down on the surface of the Moon on July 20, 1969, a moment of national pride and international sense of accomplishment. While the Soviet Union had sent more objects into space, it had not gone to the Moon. The drama of the flight and the adventure was brought straight into the living rooms of millions of people, broadcast live on radio and television around the world. For the first time, people saw the Earth as a planet as well as a global village, a distant blue-green pebble rising above an alien landscape.

Conclusion

Claude Bernard (1813–78), a medical researcher, said in a philosophical moment that "Art was I: Science is We."[4] Nowhere was this more clearly shown than in the race for space. NASA became the world's biggest supporter of scientific research in terms not only of the overall number of people directly employed or funded by grants to private and public organizations but of the scope of research and development as well. From nutritionists and home economists to theoretical physicists, from electrical engineers to librarians, from chemists to computer programmers – all were brought together in NASA. It was the biggest of Big Science. Even in 1999, when it had been reduced by almost half its peak work-force, NASA had 5,971 employees with advanced degrees (doctorates and masters) and a further 7,255 with bachelor degrees, almost all in science and engineering.

The NASA programs also helped change the image of science, especially in the United States. It was less tainted by the destructive image of the arms race and was made more American. While men speaking in foreign accents were still around, NASA's voice and image was the television news announcer Walter Cronkite and a host of boys from American coal-mining towns and prairie farms who had the "right stuff." The space program made science adventurous and glamorous rather than sedentary and esoteric. The utility of science, in its best, worst, and most Machiavellian applications, was now a part of everyday life. It had also become part of the American dream.

Essay Questions

1. What discoveries led to the formulation of the Big Bang theory?

2. How did Sputnik contribute to the development of modern science?

3. Was the Apollo program worth the money? Why or why not?

4. How was the theory of continental drift confirmed?

4. C.C. Gillispie, "Claude Bernard," *Dictionary of Scientific Biography* (New York: Charles Scribner's Sons, 1970).

CHAPTER TIMELINE

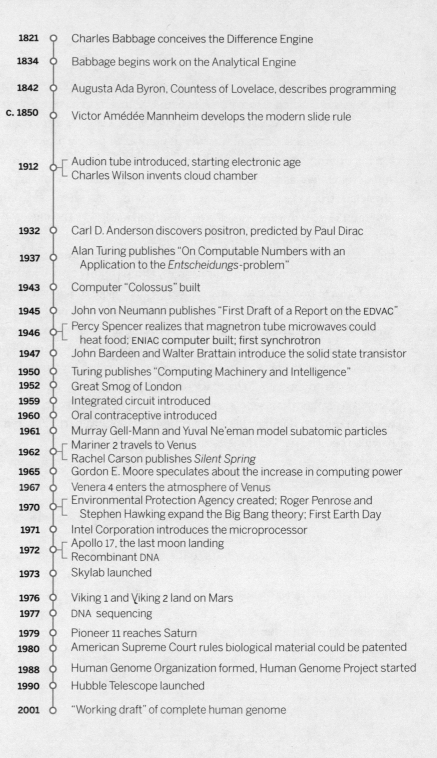

1821 Charles Babbage conceives the Difference Engine

1834 Babbage begins work on the Analytical Engine

1842 Augusta Ada Byron, Countess of Lovelace, describes programming

C. 1850 Victor Amédée Mannheim develops the modern slide rule

1912 Audion tube introduced, starting electronic age
Charles Wilson invents cloud chamber

1932 Carl D. Anderson discovers positron, predicted by Paul Dirac

1937 Alan Turing publishes "On Computable Numbers with an Application to the *Entscheidungs*-problem"

1943 Computer "Colossus" built

1945 John von Neumann publishes "First Draft of a Report on the EDVAC"

1946 Percy Spencer realizes that magnetron tube microwaves could heat food; ENIAC computer built; first synchrotron

1947 John Bardeen and Walter Brattain introduce the solid state transistor

1950 Turing publishes "Computing Machinery and Intelligence"

1952 Great Smog of London

1959 Integrated circuit introduced

1960 Oral contraceptive introduced

1961 Murray Gell-Mann and Yuval Ne'eman model subatomic particles

1962 Mariner 2 travels to Venus
Rachel Carson publishes *Silent Spring*

1965 Gordon E. Moore speculates about the increase in computing power

1967 Venera 4 enters the atmosphere of Venus

1970 Environmental Protection Agency created; Roger Penrose and Stephen Hawking expand the Big Bang theory; First Earth Day

1971 Intel Corporation introduces the microprocessor

1972 Apollo 17, the last moon landing
Recombinant DNA

1973 Skylab launched

1976 Viking 1 and Viking 2 land on Mars

1977 DNA sequencing

1979 Pioneer 11 reaches Saturn

1980 American Supreme Court rules biological material could be patented

1988 Human Genome Organization formed, Human Genome Project started

1990 Hubble Telescope launched

2001 "Working draft" of complete human genome

MAN ON THE MOON, MICROWAVE IN THE KITCHEN

12

When men last walked on the Moon in 1972, science was so entwined with the conduct of modern society in the industrialized world that it had become an indispensable component of the world it had helped to create. Most major science projects were the work of teams and networks of teams. The information being produced was so vast that not only was it impossible for a single person to keep up with developments in science generally, but scientists also had trouble keeping up with developments in their own disciplines. There were more than 20,000 scientific journals worldwide. Disciplines were splitting into subdisciplines, which in turn often split into further specialized areas. In chemistry alone the number of types of chemists had grown more than fivefold, as the disciplinary divisions in the American Chemical Society had risen from the original 5 in 1908 to 28 in 1974.

This enormous growth meant that science was no longer the exclusive domain of a philosophical elite or even of a small group of researchers/courtiers. It was an occupation, an employment choice suggested by guidance counselors, a column in the employment opportunities section of the newspaper. Increasingly, many of the people who could be classified as scientists by education and professional association were not doing research but were in charge of the many and varied jobs that required precision with scientific instruments and a knowledge of specialized systems. Or they were teaching the next generation those skills.

International Council of Science (ICSU)

Throughout the twentieth century, science was becoming a more international endeavor, even while state funding and war work seemed to be pulling it in the opposite direction. One of the most visible signs of the goal of international cooperation in science was the creation of international scientific unions. Although the new organizations traced their roots to the older multidisciplinary organizations like the Royal Society, the new groups focused on a single discipline and acted as a forum where representatives of national science bodies could come together to meet and share information. The international unions often undertook or coordinated projects, such as the IGY that had participants from several unions including the International Astronomical Union, the International Union of Geodesy and Geoscience, and the International Union of Geological Sciences. The explosion of these unions after World War II and during the Cold War speaks to scientists' desire to keep scholarly lines of communication open, even as political and ideological battles heated up. (See figure 12.1.)

In 1931 the ICSU was founded as an umbrella organization for the international unions. ICSU was the descendent of two earlier internationalist organizations, the International Research Council (1919–31) and the International Association of Academies (1899–1914). Although the scientific unions were not directly affiliated with the UN, they often worked on UN projects, particularly with UNESCO and the World Health Organization. In addition to representing science on the international stage, ICSU has worked to preserve the universality of science and defend scientific freedom. In 1998 ICSU changed its name to the International Council for Science (ICS).The ICS was expanded in 2018 to include the International Social Science Council (ISSC) and changed its name to the International Science Council or ISC.

Women in Science

In the 1970s women began to pursue scientific careers in large numbers. From the early modern period on, wives, sisters, and daughters of scientists had often made important and independent contributions to their husbands', brothers', or fathers' scientific work. Still, before the late nineteenth century, few women could choose an independent scientific career. Perhaps the most telling example of women's precarious position in science was the 1911 decision of the Académie des Sciences

12.1 INTERNATIONAL UNIONS BY DATE OF FOUNDING

1922	International Astronomical Union	Astronomy
1922	International Mathematical Union	Mathematics
1922	International Union of Geodesy and Geophysics	Geodesy and geophysics
1922	International Union of Geological Sciences	Geology
1922	International Union of History and Philosophy of Science	History of science and philosophy of science
1922	International Union of Pure and Applied Chemistry	Chemistry
1922	International Union of Pure and Applied Physics	Physics
1922	International Union of Radio Science	Radio science
1923	International Geographical Union	Geography
1925	International Union of Biological Sciences	Biology
1947	International Union of Crystallography	Crystallography
1947	International Union of Theoretical and Applied Mechanics	Mechanics
1955	International Union of Biochemistry and Molecular Biology	Biochemistry and molecular biology
1955	International Union of Physiological Sciences	Physiology
1966	International Union for Pure and Applied Biophysics	Biophysics
1968	International Union of Nutritional Sciences	Nutrition
1972	International Union of Basic and Clinical Pharmacology	Pharmacology
1976	International Union of Immunological Societies	Immunology
1982	International Union of Microbiological Societies	Microbiology
1982	International Union of Psychological Science	Psychology
1990	International Cartographic Association	Cartography
1993	International Brain Research Organization	Neuroscience
1993	International Union of Anthropological and Ethnological Sciences	Anthropology and ethnology
1993	International Union of Soil Sciences	Soil science
1996	International Union of Food Science and Technology	Food science and food technology
1996	International Union of Toxicology	Toxicology
1999	International Union for Physical and Engineering Sciences in Medicine	Medical physics
2002	International Society for Photogrammetry and Remote Sensing	Photogrammetry and remote sensing
2005	International Union for Quaternary Research	Quaternary period
2005	International Union of Forest Research Organizations	Forestry
2005	International Union of Materials Research Societies	Materials science
2011	International Sociological Association	Sociology and social sciences

to deny Marie Curie a place among its academicians. Beginning in the late nine-teenth century women began a concerted effort to break into the male preserve of science. This was part of the larger political and social feminist movement, linked to the struggle for suffrage, although it is a mistake to see the vote as the prime goal for many feminists. Education was a key component in their quest for equality, and one strategy that was used to gain greater education was the promotion of a good education to make women better mothers. Science education was central to this, since science promoted rationality and was useful. Women worked to establish a number of women's colleges in order to develop this useful science agenda. At institutions such as the "seven sisters" in the United States (Vassar, founded in 1865; Smith, 1871; Wellesley, 1875; Bryn Mawr and Baltimore, 1885; Mount Holyoke, 1888; and Barnard, 1889), women taught and learned science. The same philan-thropic women who supported these colleges also began the slow campaign to allow women into the regular universities, especially to earn PhDs in science.

Women developed branches of science that were seen as particularly suited to them, especially astronomy, psychology, anthropology, and home economics. As early as the nineteenth century women were employed as calculators in astron-omy, since they were better at the tedious mathematical calculations necessary to establish accurate star locations and would work for a fraction of men's wages. Women's assumed empathy with others made psychology (especially child psychol-ogy) and anthropology seem apt areas of study. Home economics, with its subject the science of the home, also seemed an obvious fit. Unfortunately, these proved unhappy choices. Women's services were less needed as more powerful astronomi-cal instruments and computers replaced long calculations, and the other sciences were always seen as second-class compared with physics, chemistry, or biology.

During World War II more women entered the sciences to fill positions, including university science teaching, that were vacated when many male scientists were seconded to the war effort. However, most of these new positions were not permanent, since many universities froze tenure hirings during the war, and after the war most women were expected to step down to make space for returning servicemen. Increasingly, universities and colleges, in an effort to become more prestigious, insisted that their faculty have PhDs, which far fewer women had earned. Universities also began to establish serious anti-nepotism regulations, making it impossible for the wife of a faculty member to have a job in her own right.

After Sputnik, some of the federal money earmarked for improving science in the United States did go to women, but many voiced the concern that national scholarships to women were wasted money, since girls would marry and not use the training the state had provided. The Soviet Union had many more women scientists, in part because the regime there made a conscious decision to promote science as a woman's career. Americans ridiculed these female scientists as proof of the weakness of the Soviet state.

It was not until second-wave feminists of the late 1960s and 1970s began to protest loudly the treatment of women in science that changes began to happen. Equal pay legislation in 1972, for example, forced universities and research centers to pay women the same rates as men. Anti-nepotism rules were gradually declared unconstitutional. Feminist philosophers of science critiqued the ideology of science itself, and programs were slowly put into place to encourage women to pursue scientific careers. By the end of the twentieth century women began to achieve parity with men in some sciences (most particularly the biological and medical sciences), while physics and computer science continued as male-dominated fields into the twenty-first century.

Science Produces Consumer Goods

In the postwar years, as women struggled to be taken seriously as scientists and as the status of home economics, the scientific study of the home, lost ground to other science disciplines, the relationship between the laboratory and domestic and work worlds continued to change. Science research was expected to produce useful products not just for the high-technology industries but for the average person. These products were often spinoffs of other research, such as the development of the microwave oven. The use of microwaves to heat things was a by-product of research on magnetron tubes, which were at the heart of radar-detection equipment developed during World War II. In 1946 Percy Spencer (1894–1970) was working for the Raytheon Corporation on ways to improve the manufacturing of magnetron tubes when he noticed that they heated things. In tests, he popped popcorn and exploded an egg by placing them close to a working magnetron tube. He applied for a patent (1 of 120 he received) for a microwave cooker, and in 1947 the first "Radarange" was produced. It was about the size and weight of a refrigerator and required plumbing for the cooling system. However, by 1965 domestic countertop models were introduced and are now a standard kitchen appliance in the industrial world.

Another example of how developments in research were folded back into scientific work to produce consumer goods occurred in the electronics industry. Around the beginning of the twentieth century, a host of scientists and inventors such as Thomas Edison (1847–1931) and Nikola Tesla (1856–1943) had bridged the gap between theoretical and experimental science and application. Their inventions, the teams they created, and the companies that developed their work brought electrification to industry and the home. The electrical industries, including the telecommunications sector, required workers who were part engineer, part scientist, and part technician. These people could work with the basic principles of electricity, but they were generally asking the question "what can I make this do?" rather than the question "how does the universe function so this is possible?" This is sometimes called "applied science," but there is no clear line between applied and pure science, either historically or institutionally, particularly when the products of applied science were then used in fundamental research in the form of new instruments.

Two discoveries in particular combined to change the practice of science. They were the development of computing and the introduction of the solid-state transistor. Independently, each was an important innovation, but together they transformed industrial society. As Big Science coordinated the work of larger and larger teams, computers offered the technology to manage both the equipment of Big Science and the oceans of information such projects generated.

Computing

The origins of computing were analogous with the development of mathematics. The use of tally bones dates from around 100,000 BCE, while the first successful mechanical calculating device was the abacus. Some historians identify a kind of counting table that worked on the same principle as the more familiar frame abacus as far back as 3,000 BCE in Babylonia. The Babylonians, who were excellent mathematicians, made clay tablets listing tables of important numerical information as early as 2,300 BCE. By about 400 BCE the abacus was used around the Mediterranean. Whether through independent development or trade contact, the abacus was developed in China in its modern form of beads on rods or wires sometime after 600 CE.

The abacus was useful for rapid calculations involving basic operations, and it remained the primary calculating device for most of the world until the twentieth century. However, it was less useful for the kinds of questions examined by natural philosophers and mathematicians by the seventeenth century. Problems in trigonometry involving sines and tangents and greater facility with complex calculations such as square roots were either very time-consuming or impossible using an abacus. John Napier (1550–1617) turned his mind to the theory of calculation in 1594. His work was on algorithms, or methods of calculation. In 1614 he published *Mirifici logarithmorum canons descriptio* (*Descriptions of the Marvelous Rule of Logarithms*), which contained his tables, the rules for their use especially for trigonometry, and the principles of logarithmic construction. The tables had taken him 20 years to complete, but they were so well done that no serious revision was considered necessary for over 100 years.

Napier's work was expanded by other mathematicians who sought to transform logarithmic tables into mechanical form. Both Edmund Gunter (1581–1626) and William Oughtred (1574–1660) produced types of slide rules that used logarithms; Victor Amédée Mannheim (1831–1906) developed the modern slide rule about 1850.

Mechanical calculators, usually based on some system of gears, also have a long history. Descriptions run back into antiquity, with a technological overlap between mechanical calculating devices and other machines such as clocks and astrolabes. In the early modern period Wilhelm Schickard (1592–1632), an astronomer and cartographer, described a mechanical calculator to Kepler in a letter in 1623, but it was not until 1642 that Blaise Pascal presented a working mechanical calculator.

One of the constraints on calculators was the precision necessary to manufacture the component parts. In the nineteenth century precision fabrication reached new levels of quality, so when Charles Babbage conceived of his Difference Engine in 1821, the technology was available to construct his device. The Difference Engine was designed to calculate and print mathematical tables using polynomial functions, all without the necessity of human intervention. (See figure 12.2.) It was financed in large part by the British government, which paid Babbage £17,500, a huge sum of money. The government hoped that the Difference Engine would be useful in such areas as navigation, astronomy, and calendars. Early on it appeared that the investment might offer some return. Babbage completed enough of the Engine so that in 1827 it was used to calculate tables of logarithms, but instead of finishing the work, he continued adding to its design, requiring it to be partially dismantled and thus delaying its completion. Then he abruptly abandoned the enterprise altogether in 1834 to begin work on a new type of calculating device he

B. H. Babbage, del.

12.2 BABBAGE'S DIFFERENCE ENGINE

A small portion of Babbage's Difference Engine No. 1 from his autobiography, *Passages from the Life of a Philosopher* (1864).

called the Analytical Engine. This new device could be programmed by using punched cards so that it was a general purpose machine, rather than simply a kind of mechanized adding machine. The British government, having received nothing for their earlier investment, refused to fund the new machine regardless of Babbage's claims for its amazing potential. Without funding, the Analytical Engine was never built.

The idea of a calculator whose functions could be changed, or programmed, to suit the circumstances was a crucial step toward the creation of computers. Babbage had conceived of the Analytical Engine storing results (the Store) separately from the part of the machine that did the calculating (the Mill). The idea of punched cards to control the engine had been borrowed from the textile industry, where Joseph-Marie Jacquard (1752–1834) had introduced a system of punched cards to control the patterns woven by looms. The best explanation of the engine's operations came not from Babbage but from Augusta Ada Byron, Countess of Lovelace (1815–52). The daughter of the poet Lord Byron, Ada met Babbage in 1833 and asked him to teach her mathematics. In 1842 Babbage asked her to translate a French account of the technical aspects of the Analytical Engine, but her commentary was longer than the original document. In 1843 she wrote about a special class of calculus functions – this was one of the first descriptions of programming. Historians disagree about Ada's mathematical abilities and her role in Babbage's work. While she is sometimes called the first "computer programmer," this is somewhat ahistorical. Babbage and his son, who assisted him on the development of the Analytical Engine, were both conversant with the theory and practice of the engine's operations before she was involved in the project. Nonetheless, Ada's contribution in an age when women were actively discouraged from such intellectual activity was remarkable. In honor of her work, in 1979 a programming language developed for the American Department of Defense was named Ada.

Although Babbage never completed his Analytical Engine, the abandoned Difference Engine No. 2 was built at the Science Museum of London in 1991. Weighing two-and-a-half tons and containing more than 4,000 moving parts, it worked just as Babbage had said it would.

Although a small number of difference engines were built, including a motorized version exhibited by George Grant in 1876 in Philadelphia, they were not widely used because there was little market for them. They were generally too expensive for business applications and not versatile enough for mathematical or scientific work. It was not until the introduction of vacuum tubes that offered high-speed switching and the technical demands of World War II that the development of electronic calculating devices moved beyond the theoretical complexity of Babbage's mechanical designs. The Ultra program allowed the British to decrypt German codes, but it took too long. In 1943 Tommy Flowers (1905–98) was sent to Bletchley Park, home of Britain's secret team of code breakers, to build an electronic code breaker. The result was Colossus, which gave the Allies a huge advantage over German forces during the invasion of Europe. Despite its importance, little remains of the Colossus project. The original machines were destroyed in 1946, and even their existence was kept secret until 1970.

On the other side of the Atlantic John Mauchly (1907–80) and Presper Eckert (1919–95) built ENIAC (Electronic Numerical Integrator and Computer). The suggestion that a fully electronic calculator (except for input and output elements) could be developed was made around 1942. One of the main motivations for such a device was the increasing complexity of long-range artillery, which could fire at targets far out of sight. Ballistics tables and firing solutions were so complex and time-consuming that the most complex might take hours to compute by hand. So the American Army Ordnance Corps funded the development of ENIAC, spending almost $500,000 on the project. Mauchly and Eckert worked on and oversaw the construction of the massive machine. It weighed over 30 tons, contained 19,000 vacuum tubes, and consumed almost 200 kilowatts per hour of electrical power to run both the electronics and the cooling system that kept it from burning out.

ENIAC was completed in 1946 at the Moore School of Electrical Engineering at the University of Pennsylvania. After it was tested, it was taken apart and transferred to the Aberdeen Proving Grounds, where it went into general operation in 1947. Although ENIAC was not operational during the war, it was used for a number of other projects, including weather forecasting, satellite trajectories and orbits, and nuclear weapons programs. Colossus had been the first electronic computer, but it was from ENIAC that the computer industry sprang.

Computing Theory: Turing and von Neumann

While computers were at some level an engineering problem, the theory of computers was greatly affected by two men: John von Neumann (1903–57) and Alan Turing (1912–54). Von Neumann was a mathematical prodigy who turned his hand to a wide range of problems from game theory to subatomic physics. He arrived in the United States from Hungary in 1930 and was one of the first people to be appointed to the Institute for Advanced Studies when it was founded in 1933. During the war, von Neumann was a consultant to the Manhattan Project and to the Ordnance Corps, which brought him into contact with the ENIAC group. Turing was also a prodigy, with interests in science and mathematics. His wartime experience was at Bletchley Park, where his work on Project Ultra broke the German Enigma code. Like von Neumann, his work gave him access to actual computing devices.

In 1935 Turing began exploring the theoretical possibilities of computation. In his 1937 paper "On Computable Numbers with an Application to the *Entscheidungs*-problem," he speculated about a machine that could carry out calculations based on a finite table of operations and reading or deleting a series of instructions on a paper tape. This theoretical device, later called a "Turing machine," offered a number of theoretical ideas that underpin modern computers. After the war Turing was invited to join the National Physics Laboratory and asked to build a computer. His plans for the Automatic Computing Engine were never carried out, however, and he moved to the University of Manchester to work on the Manchester Automatic Digital Machine. In 1950 he published "Computing Machinery and Intelligence" in the journal *Mind*. It outlined many important issues in computing, speculated on the relationship between computers and human thought, and outlined the "Turing test," which argued that machine intelligence could be evaluated by observation of the interaction between a computer and a human. If a human asking questions of a hidden respondent could not discern whether the respondent was a computer or another human, then the computer had likely achieved a level of intelligence equivalent to human intelligence.

Von Neumann's experiences with computers likewise led him to see them as offering a powerful tool. In his "First Draft of a Report on the EDVAC" (1945), he described a general purpose, stored program computer. Many of his ideas were translated into actual computer equipment, and, as a consultant to IBM he contributed to the development of commercial computers in the 1950s. In 1953 the first production model, the IBM 701, was marketed. Nineteen were sold, mostly to aerospace contractors and the government. One of its main uses was, in fact, the development of nuclear weapons. In these two realms – computing and nuclear weapons – von Neumann had a profound influence, which he used to try to persuade the American government to take military action against the Soviet Union. Partly based on his anti-communist political views and partly on ideas that he had developed when he created the economic/mathematical field of game theory, von Neumann advocated using the atomic bomb on the Soviet Union before it could develop its own nuclear weapons.

Both von Neumann and Turing died young. When Turing was arrested for having a homosexual relationship (illegal at the time), the British government removed his security clearance, effectively driving him out of government work amid Cold War paranoia. He committed suicide in 1954. Von Neumann died of cancer only three years later. While many people worked on computers, these two were leaders in the theoretical and conceptual aspects that made the electronic computer the most versatile device ever created.

The Solid State Transistor

Turning computing machines from gigantic monsters full of temperamental vacuum tubes that had to be tended by a phalanx of technicians into something more manageable and affordable required the introduction of a new technology. That technology, which had started at almost the same moment as ENIAC began to calculate ballistic tables for the American army, was the solid state transistor.

The solid state transistor originated as a solution to a fundamental problem in electrical communication. As the distance an electrical impulse (the signal, whether dots and dashes, voice, or later other kinds of information) had to travel increased, it progressively weakened. This was a serious problem for long-distance telephones, so many people worked to figure out how to boost the signal strength

12.3 SOLID STATE AMPLIFIER

without distorting the signal. It was partly solved by the introduction in 1912 of the Audion tube and its many descendants. For telephone companies with growing numbers of subscribers, the tube solution created its own technical problems, since the amplifiers were bulky, used a great deal of power, and could knock out service for long periods by burning out.

In 1936 the physicist William Shockley (1910–89) was hired by Bell Laboratories and, with his knowledge of quantum mechanics, was asked to look for ways to create a reliable solid state device to replace the old system. His preliminary efforts were unsuccessful, and then the war interrupted work. When he returned to Bell Laboratories in 1945, he returned to the solid state question, but another design failed. The project was then turned over to John Bardeen (1908–91) and Walter Brattain (1902–87), two scientists working in Shockley's lab. They came up with a new approach. Taking a small piece of germanium, some gold foil, and a paper clip as a spring, they were able to amplify an electrical signal almost 100 times its original strength. (See figure 12.3.)

On December 23, 1947, Bardeen and Brattain demonstrated their invention to Bell executives. It was a major breakthrough, although beyond its application to telephone amplification, no one really knew what it would mean for the electronics industry. Further, Shockley was both pleased with the success of his research group and angry that he had not been the inventor after so many years of effort. He went on to develop a different form of transistor that replaced the fine wires used to channel the signal through the semiconducting material of the Bardeen/Brattain design (called "point contacts") with a more robust system based on "junctions" or boundaries in the semiconducting material itself.

In the beginning the transistor found limited use. In 1952 the first commercial product using the new technology was a hearing aid. Large electronics companies such as General Electric, Philco, and RCA had too much invested in tube technology to simply change to the new transistors. The first mass market effort was the production of junction transistors by Texas Instruments for a portable radio in 1954. In 1958 Jack Kilby (1923–2005) at Texas Instruments put different electronic components on the same piece of semiconductor, while in 1959 Robert Noyce (1927–90) at Fairchild Semiconductors worked out a way to link such components. Between them, the two men created the integrated circuit. Texas Instruments got out of the radio market, and a small Japanese company, Sony, jumped in. Transistor radios, relatively inexpensive and small enough to fit in a shirt pocket, hit the market just as postwar prosperity and the Baby Boom made electronic entertainment big business. By the middle of the 1960s Sony was manufacturing television sets using transistor technology, and Japan was starting to dominate the consumer electronics industry.

The application of the transistor affected various aspects of computer technology, and in 1964 the Comcor Company produced the ci 5000, the first fully transistorized, general purpose computer. The biggest breakthrough after the integrated circuit was the 1971 development by Ted Hoff (1937–) at Intel Corporation of the microprocessor, which placed all the logical elements necessary for computing on a single chip. The microprocessor controlled the states (the on/off conditions) that made programming and calculation possible, as well as handling the input and output of signals. After that, electronic technology exploded. Microprocessors were made not just for computers but were put into almost anything that used electricity, from children's toys to elevators, automobiles, and pacemakers. What had begun as a question of signal strength became the backbone of modern industrial society, controlling the flow of information, monitoring the environment, and containing the massive amounts of information necessary to keep all the computer-controlled systems operating.

The microchip revolution was predicted in 1965 by Gordon E. Moore (1929–), the head of research at Fairchild Semiconductors. He observed that the complexity of integrated circuits had roughly doubled each year since 1959; using that basic rate, he predicted that by 1975 there would be 65,000 transistor chips on the market. Although that took a bit longer – it wasn't until 1981 that chips hit the 65,000 mark – his observation about the rate of increase was labeled "Moore's Law" and has been a kind of rule of thumb for microprocessors ever since. While superchips, at the very limits of the materials and manufacturing systems, are already in the planning and development stage, it is not the hardware alone but the software that ultimately makes the computer useful.

The introduction of computers, both large and small, transformed scientific practice in almost all disciplines. Anything requiring many mathematical operations or large data sets, such as astronomy and subatomic physics, could be handled more quickly and accurately by computers. However, computers could also be used to control equipment, so everything from microscopes to electrophoresis became mechanized. The utility of computing power was so significant that scientists around the world also worked on systems to connect computers and to gain access to computer power at distant locations, giving birth to computer networks and ultimately to the internet.

The Pill: Science and Gender Relations

While radios, televisions, electronic game consoles, and home computers made inexpensive by transistor technology were entertaining the population and providing a means of stimulating demand for more consumer products, another branch of science was about to change gender relations. This change took place in the bedrooms of the industrial world. In the struggle for gender equality, science had been used as a justification both for and against women's rights. Much of the debate hinged on reproduction. For generations, many people (both men and women) had argued that, since childbearing was the highest purpose for women, anything that took women away from that goal (such as education, science research, or work outside the home) was not just a bad idea but contravened nature. A number of women's rights advocates had worked on birth control as a key both to women's independence and to a better life for everyone. To make matters more complicated, some birth control advocates were connected to the eugenics movement in the early twentieth century, which sought a differential birth-rate for the "fit" and "unfit." Originally, birth control depended on barrier methods such as condoms and cervical coverings or on less reliable ovulation timing methods such as the rhythm method. This all changed in 1951 when Carl Djerassi (1923–2015), while conducting research on the hormonal control of reproduction with a team at the University of Mexico, showed that ovulation could be controlled by administering the hormone progestin orally. An oral contraceptive became a possibility.

The search for a reliable form of hormonal contraceptive was continued by Gregory Pincus (1903–67) and Min Chueh Chang (1908–91) at the Worcester Foundation for Experimental Biology. Margaret Sanger (1879–1966), an early birth control advocate, introduced Katherine Dexter McCormick (1875–1967), heir to the International Harvester fortune, to the researchers, and McCormick agreed to help fund their work. In all, she contributed some $3 million to research on oral birth control. In 1956 large-scale human trials were conducted by John Rock (1890–1984), and in 1960 the American Food and Drug Administration granted approval for the sale of oral contraceptives. Although there were some problems with the initial drug (it was found to contain about ten times more hormone than was needed for contraception) and some jurisdictions were slow to legalize this new contraceptive, it had a profound effect on Western society. Because of this invention, relationships between the sexes changed radically and made the sexual revolution of the 1960s possible.

Following another line of inquiry, in 1961 Jack Lippes (1924–) introduced the Lippesloop, an IUD (intrauterine device). The IUD was a small inert object (usually made of plastic) inserted in the uterus to prevent fertilization, although the exact reason why it works is still not completely understood. Various forms of IUD may have existed from ancient times, but their modern development began in 1909 with work by Ernst Grafenberg (1881–1957). It was initially seen as a safe, low-cost method of birth control, which was also under the control of women. IUDs were popular, but their use in North America plunged after the A.H. Robins Company was sued over problems with the Dalkon Shield. The Dalkon Shield had been developed by Hugh J. Davis, a physician at Johns Hopkins Medical School during the late 1960s. Design problems were linked to pelvic inflammatory disease, uterine damage, sterility, and a number of deaths. Almost 400,000 claims were brought against the manufacturer, which stopped sales in the United States in 1974 but not in other parts of the world until several years later. Since IUDs were not drugs, there was little control over their development, testing, or marketing. The legal case highlighted the difficulty of using scientific evidence in legal matters, as both sides brought in scientific experts to argue their case. According to some, the Dalkon Shield was safe, while others presented evidence suggesting it was dangerous. The problems with the Dalkon Shield affected the perception of IUDs in general and, especially in the industrialized world, their use dropped dramatically.

Exploring Space

The Age of Aquarius, the Woodstock generation, and the sexual revolution of the late 1960s and 1970s were the creation of modern contraceptives. Science shaped this new age in a very real way, but it also raised fears and disenchantment.

Concerns about nuclear Armageddon, the war in Vietnam, and the close ties between scientists and the military contributed to a sense of social unease with science and technology, especially in the United States. Even the flagship of American science and technology, the Apollo program, ended in 1972 when Apollo 17 became the last manned mission to the Moon. In 1973, Skylab was launched, but it largely failed to capture the imagination of the public. Many scientists also questioned the utility of the Skylab missions, although they did help provide information about the long-term effects of zero-gravity environments on the human body.

The exploration of space was given over to probes, which made more sense technologically, financially, and scientifically but lacked the romance of a human explorer, a Columbus or Neil Armstrong. Mariner 2 traveled to Venus in 1962, Venera 4 made it into Venus's atmosphere in 1967, and Venera 9 broadcast television pictures from the planet's surface in 1975. Venus, once thought to be a water world, turned out to be a hell of high temperatures, carbon dioxide atmosphere, and acid rain.

Mars, Earth's other close neighbor, was also the target of many probes. In 1971 two Soviet and one American spacecraft went into orbit around the red planet. At first the images they sent back were disappointing, revealing little about the surface of the planet, but that turned out to be due to a planet-wide dust storm. Over time, the Mars satellites sent back large amounts of information. Several attempts were made to land probes on the surface, but it was not until 1976 that Viking 1 and 2 landed and remained operational. While the old science fiction idea of a Martian civilization had long since disappeared, there were hopes that life in the form of microbes might be found, but the probes found nothing. To date, no trace of life on Mars has been found, but there remain theories about traces of life in the polar regions where there is more moisture, and recently running water seems to have been detected, with the Mars Exploration Rovers sending back evidence of water in 2015.

Also in 1974 Mariner 10 reached the vicinity of Mercury, showing it to be a lump of baked rock with a lunar-like surface of craters. Probes to the outer planets started with Pioneer 10, which flew by Jupiter in 1973, sending back information about the giant planet's magnetic fields and moons, before heading out into deep space beyond the orbit of the furthermost planet in the solar system. Pioneer 11 reached Saturn in 1979, followed by Voyager 1 and 2 in 1980 and 1981. They sent back spectacular images of Saturn's rings and revealed the presence of previously undetected moons. While Voyager 1 headed into deep space, Voyager 2 continued on to Uranus, flying past it in 1986.

Physics: Particles within Particles

As astronomers were rapidly learning a tremendous amount about outer space, physicists investigated the minuscule. Since the era of Rutherford, physicists had been looking at the structure of atomic particles such as photons, electrons, and neutrons. They had been able to study the behavior of particles in a number of ways, but the 1912 invention of the "cloud chamber" by Charles Wilson (1869–1959), which used water vapor (and later alcohol) in a small enclosed chamber to track particles, made the process much more versatile. Particles entering the chamber caused a small amount of condensation along the path they followed. Because of the behavior of the particles, these paths curved or made coils, allowing scientists to calculate energy, charge, and life span.

In 1932 Carl D. Anderson (1905–91) used a cloud chamber to study cosmic rays, attempting to determine if they were particles or waves. He found that the rays could be deflected by a strong magnetic field, indicating a particle nature. At the same time he noticed a track for another particle that was the same as that for an electron, but it curved in the opposite direction, indicating that it had the opposite charge. This particle was an "anti-electron" or positron (from "positive electron"). Its existence had been predicted several years earlier by the theoretical physicist Paul A.M. Dirac, who suggested that every subatomic particle would have a corresponding antiparticle.

Other kinds of particles were discovered from cosmic ray studies, such as two "mu-mesons" (shortened to "muon") found by Anderson in 1936. These particles had opposite charges and were much more massive than electrons but lighter than protons. This was followed by the discovery of three forms (positive, negative, and neutral) of "pimeson" (or "pion") in 1947 and in 1952 of the "lambda" particle, which had no charge of its own but was found to break down or decay into a proton and a negative pion.

The particles that made up atoms were literally and theoretically coming apart, but studying the bits was a technical challenge. To study the particles it was necessary to accelerate them, smack them into a target (other particles), and see what came apart. This was accomplished with machines known (not surprisingly) as accelerators, invented in 1928 by John Cockcroft (1897–1967) and Ernest Walton (1903–95). These were linear accelerators that projected the particles down long straight tubes or tunnels. While useful, they had limitations, and in 1931 Ernest Lawrence developed the first working cyclotron, which used a ring of magnets to accelerate particles. Cyclotrons confirmed Einstein's theory of relativity since, as the electrons were accelerated toward the speed of light, they gained mass. This became apparent because the magnetic field required to keep the electrons moving around the cyclotron had to be greatly increased as they sped up. But to break apart the atomic particles in order to try and find the range of subatomic particles predicted by theorists required huge amounts of energy, and eventually scientists built a new class of accelerators that synchronized the increase in magnetic field needed as the particle circled through it. These machines were called "synchrotrons"; the first one was created in 1946. Further refinements also allowed experimenters to link accelerators, so that in 1983 Fermilab created an accelerator that used a linear accelerator, a 152-meter diameter synchrotron, and a four-mile diameter synchrotron.

In addition to the new accelerators a more sensitive method of recording the particle tracks was developed. In the early 1950s Donald A. Glaser (1926–2013) created the "bubble chamber," which worked on the same principle as the cloud chamber but used liquid hydrogen heated to just below its boiling point rather than water or alcohol vapor. When charged particles passed through the hydrogen, they added just enough energy to cause it to boil and release bubbles. By measuring the path and the number of bubbles produced by the passage of a particle, information about its mass and velocity could be determined.

With all this new equipment, a host of subatomic particles were found. In addition to the old proton and electron were the muon, kaon, gluon, sigma, xi, lambda, eta, and a host of neutrinos. In 1961 Murray Gell-Mann (1929–2019) and Yuval Ne'eman (1925–2006) independently created a theoretical model to classify the many particles into families, somewhat like Mendeleev's periodic table of elements. Yet there was a mystery about the apparent creation of large numbers of subatomic particles. A group of particles known as "hadrons" seemed to be almost limitless in number, an odd finding for something that was supposed to be more fundamental or elementary than the large bits of the atom. In 1964 Gell-Mann and George Zweig (1937–) theorized that there was a parallel between hadrons and molecules. There were, of course, a huge number of different molecules, but only a small and limited number of elemental atoms. The hadron particles, Gell-Mann and Zweig argued, were to the subatomic world what molecules were to the atomic world. They predicted the existence of truly elementary particles that Gell-Mann named "quarks," borrowed from a line in James Joyce's novel *Finnegans Wake*. In this model, each proton or neutron was made up of three quarks plus "gluons" that held the quarks together.

Physics: The Very Big

The very big and the very small had to intersect, because the creation of all these particles had to come from some source. According to the Big Bang theory, before the beginning of the universe all energy and matter were confined to a point with no dimensions. It was very hot, and the energy of the universe was so high that there were no particles and all forces were essentially unified or completely symmetrical. When the universe came into being, it started to cool and parts started to differentiate. This differentiation, or asymmetry, led to the creation of matter, building up from the subatomic realm, which clumped together in various ways to give us the universe as we now see it. This was the Big Bang theory as outlined by Friedmann and expanded in 1970 by Roger Penrose (1931–) and Stephen Hawking (1942–2018).

Although there are still many points of debate about the details of the origin of the universe, subatomic physics and astrophysics have tended to confirm the Big Bang model. The study of the very big (astronomy) and the very small (subatomic physics) also come together in theoretical efforts to create a "grand unified theory," or GUT, that will provide the mathematical model for all four forces (electromagnetic, gravitational, weak nuclear, and strong nuclear), in effect rolling particles back to the original symmetrical condition at the Big Bang. Experimental results have shown that the distinction between the electromagnetic force and the weak nuclear force (which is responsible for nuclear decay, as opposed to the strong force that holds the atomic nucleus together) disappears at very high energy levels, so the two forces look like just one force.

In the 1970s a new tool was conceived to help look for evidence of the early state and evolution of the universe as well as to broaden optical astronomy. A space-based telescope was first proposed as a joint project between the European Space Agency and NASA. The Hubble Space Telescope, named after Edwin Hubble who had first observed the expansion of the universe, was launched into orbit aboard the space shuttle Discovery on April 25, 1990. It provides images free of the distortion of the atmosphere, although a flaw in the optical system had to be corrected in space in 1993. The Hubble Telescope provides huge amounts of data, about 14,000 megabytes daily. (See figure 12.4.) The Eagle Nebula is one of the most famous pictures taken by the Hubble Telescope.

12.4 EAGLE NEBULA (NASA)

"Appearing like a winged fairy-tale creature poised on a pedestal, this object is actually a billowing tower of cold gas and dust rising from ... the Eagle Nebula. The soaring tower is 9.5 light-years or about 90 trillion kilometers high, about twice the distance from our Sun to the nearest star. Stars in the Eagle Nebula are born in clouds of cold hydrogen gas that reside in chaotic neighborhoods, where energy from young stars sculpts fantasy-like landscapes in the gas. The tower may be a giant incubator for those newborn stars. A torrent of ultraviolent light from a band of massive, hot, young stars (off the top of the image) is eroding the pillar."

Source: Yury Dmitrienko / Shutterstock.com. Caption from NASA, ESA, and The Hubble Heritage Team (STScI/AURA), https://esahubble.org/images/heic0506b/.

Ecology and the Environment

Although subatomic particles and the origin of the universe were brought to the attention of the public by people such as Carl Sagan and his television program *Cosmos*, and Stephen Hawking's best-selling book *A Brief History of Time* (1988), public interest definitely shifted back to the home planet during the 1970s and 1980s.

One of the key reasons for this was the rise of the ecological movement. What we think of today as environmentalism is not a new development but can be traced back to efforts in the nineteenth century to conserve wilderness areas and to limit water and air pollution. Even earlier, societies from the ancient Egyptians on had laws about water use, garbage collection, and human waste disposal, and these laws existed in various forms wherever urban centers appeared. Yet the scientific study of ecology came relatively late. Biology tended to look first at individual organisms and later at genetics, rather than the organism in the environment. Similarly, chemists were focused on the controlled use of chemicals, not on the effects of uncontrolled chemicals, while physicians were trained to look for particular pathogens and were less aware of the interaction of people with the environment.

In 1962 the inseparable relationship between people and their environment was brought forcibly before the public when Rachel Carson (1907–64) published *Silent Spring*. Aimed at a popular audience, her book introduced many to the problem of environmental pollution, in particular, the problem of pesticide use. Carson pictured nature as an interlocked system rather than a series of independent components that could be treated individually. A farmer spraying DDT on crops to control insects was not just exposing the crops to the chemical but also leaving the chemicals in the ground where they could be absorbed into the tissues of the plants they were meant to protect and by other crops long after the spraying ended. These levels could build up, remaining in the soil for many years.

But the persistence of pesticides was not the worst part. The chemicals worked their way into the food chain, accumulating in higher and higher levels in the tissues of those animals at the top of the predator cycle. Plants contaminated insects, which in turn were eaten by songbirds. The concentration of toxic chemicals could grow so high in the songbirds that they would die or be unable to reproduce. Their death was the inspiration for the title of Carson's book. If action was not taken, argued Carson, there could truly be a silent spring. The death of songbirds was one of a litany of dangerous effects caused by pesticides, and Carson presented numerous stories about the outright abuse of chemical agents, a lack of scientific study of the long-term effects of pesticides by either government or manufacturers, and inadvertent or deliberate cover-ups of mismanagement of chemical agents. For instance, the pesticide Aldrin was sprayed over parts of Detroit by Michigan state officials to control Japanese beetle. When the spray planes began to fly with no prior warning, concerned citizens flooded city officials and the Federal Aviation Agency with calls. Citizens were told that the spraying was completely safe, even though the American Public Health Service

CONNECTIONS

Earth Day and
the Rise of
Environmentalism

While scientists and politicians had been concerned about clean air and water for generations, and while various pieces of anti-nuisance and anti-pollution legislation had been passed in Europe beginning in the nineteenth century, it was not until the 1960s that popular action forced governments and businesses to act decisively on these issues. One of the popular calls to action was the creation of Earth Day, a day for citizens to advocate for a cleaner and more sustainable environment. The original idea for Earth Day came from two sources: peace and environmental activist John McConnell (1915–2012) proposed a day to celebrate peace and ecology at a UNESCO conference in San Francisco in 1969; and a month later US Senator Gaylord Nelson (1916–2005) proposed an Earth Day at an environmental teach-in after witnessing the damage caused by an oil spill on the coast of Santa Barbara. Much of the early organizing for Earth Day was done by environmental activist Denis Hayes (1944–), and the first Earth Day was celebrated on April 22, 1970, with thousands of participants at 2,000 colleges and universities and 10,000 primary and secondary schools across the United States.

Earth Day was the result of the growing global concern about human degradation of the environment. One of the biggest concerns for environmental activists in this period was smog. Combining the words "smoky" and "fog," smog was seen to be obscuring the sky over every major city in the world. In Britain "killer fogs" or "pea soupers" were

and the Fish and Wildlife Service both had published reports of the toxicity of Aldrin. Large numbers of birds died, while animals and humans exposed to the chemical got sick.

Carson's research struck a responsive chord with a public increasingly disenchanted with the changes brought about by science and fearful of its power. By 1969 there was enough popular support for government action on environmental concerns for the United States to pass the National Environmental Policy Act; in 1970 the Environmental Protection Agency (EPA) was founded to enforce that legislation as well as other laws such as the Clean Air Act. In the same year the created by natural fog combining with toxic sulfur dioxide and soot from coal fires to produce smog so thick it was almost opaque and could actually kill people, especially the elderly or those with breathing problems. The Great Smog of 1952 in London made more than 100,000 people sick and perhaps caused as many as 12,000 deaths. This crisis led directly to the British Clean Air Act of 1956, which limited the domestic use of coal in British cities.

Another aspect of environmental degradation was brought to public attention by Rachel Carson. When she published *Silent Spring* (1962) to make people aware of the environmental damage being caused by pesticides, particularly DDT, American citizens cried out for government legislation to protect people and nature. US public concern about air pollution, water pollution, and pesticides led to the passing of the National Environmental Policy Act of 1969 and the creation of the EPA in 1970. The EPA was given the mandate to enforce both new rules and a range of environmental laws that were already in place.

The concern that the environment was not just a national issue but a global one was heightened by Paul R. Ehrlich's book *The Population Bomb* (1968) and the Club of Rome report *The Limits to Growth* (1972), both of which made dire predictions about the potential for disaster caused by the growth of global populations and environmental collapse. In 1971 Greenpeace was founded, originally to protest against nuclear tests and radioactive fallout, but quickly becoming concerned with wider environmental issues. In 1972 the UN Environment Program began in an attempt to coordinate international efforts. There have been some successes from these environmental efforts: DDT was banned, most industrial countries began to produce less smog, and whaling was outlawed, as were ozone-depleting chemicals.

Scientists have been important players in these environmental movements. But they could not have made these changes without the support and participation of citizens around the world. The challenge was to engage the public and give them the tools to fight for legislative and regulatory change.

EPA started operating, the first Earth Day was celebrated. Based on the teach-ins of the movement protesting the war in Vietnam, it was designed to bring the issues of pollution to the public. Such public interest also helped legislators ban the pesticide DDT in 1972.

The dangers of chemical contamination were further thrust into public view when the Love Canal story broke in 1978. Love Canal was a residential development in Niagara Falls, New York, that had been built on or beside a chemical waste dump that the Hooker Chemical Company operated from 1920 to 1953. In 1976 residents began to complain about bad odors and oozing black or brown sludge. By

1978 a major battle over the safety of the neighborhood was underway. Toxic chemicals including dioxin, known to be a powerful carcinogen, were found on the site, but the degree of danger was very much at issue, with experts lining up on both sides of the case. The federal government stepped in when President Jimmy Carter declared the site a Federal Emergency zone, and the residents were relocated. Claims and counter-claims about the health risks continue to be made to this day, with some environmentalists and scientists claiming that the danger was very great, linking such environmental damage to cancer, miscarriage, and declines in sperm counts not just in Love Canal but worldwide. Other scientists have argued that there was no sound evidence of increased health problems, again showing that scientists could be found to support opposing positions. Such controversies tended to confirm the public's lack of faith in an "objective" scientific answer about dangers that appeared to be created by scientists in the first place.

The environmental movement represents an intersection of scientific exploration and public participation. At one end of the spectrum are the eco-warriors for whom any scientific information is suspect because science is unnatural and scientists have helped make the destruction of nature possible. Therefore, only direct action to stop environmental destruction will do any good. At the opposite end are the pro-science groups and individuals who range from scientific optimists, who believe that science will solve many environmental problems, to the debunkers, who use their scientific skills to counter any claims of ecological damage they see as being based on bad science. In between are both nongovernmental organizations, such as the Sierra Club, Pollution Probe, and the Club of Rome, and governmental organizations, such as the EPA and the World Health Organization, that attempt to use science to indicate where problems exist and to support actions to solve them.

DNA and the Human Genome Project (HGP)

Human control of the environment became even more direct during the postwar period because of genetic research. The discovery of the structure of DNA offered the possibility of directly manipulating the control mechanism of cell activity, thereby giving us the ability to change and ultimately create new kinds of life forms. It has commercial possibilities in genetically altered crops and food, as well as invoking the recurring specter of modern eugenics made possible by the direct manipulation of humans.

Since Watson and Crick uncovered the structure of DNA, thousands of research hours have been poured into studying its function and working out methods to manipulate it to give an organism new characteristics. In many ways uncovering the structure of DNA was like being given the owner's manual to an expensive car and finding that it was written in code. Certain things were obvious from observation, but the details of how DNA worked still needed to be figured out. At the heart of the decoding effort were the base-pairs, the a-t, c-g combinations that made up the rungs of the ladder. Somehow, the sequence of bases controlled cell function. In 1961 Sydney Brenner (1927–2019), Francis Crick, and their team argued that the bases could be read in strings of three (such as t-t-g and the complementary a-a-c), which they call "codons." Codons allowed RNA (ribonucleic acid, a kind of molecular robot) to control the actual production of proteins. They called the molecules "transfer RNA" or tRNA.

In the same year, Brenner, François Jacob (1920–2013), and Matthew Meselson (1930–) discovered messenger RNA (mRNA), which carries the pattern of part of the DNA to the ribosomes, the site of protein synthesis in the cell. With the system to transfer information from DNA to protein production, the method of cellular control was revealed. (See figure 12.5.)

This discovery opened up the possibility of interacting with DNA, first by identifying what parts of the huge molecule were responsible for what enzyme or protein chains, and then by getting DNA to do what was wanted. The isolation of restriction enzymes that could cut DNA at specific sites meant that parts of DNA could be looked at separately. The first of these molecular scalpels was identified by Hamilton Smith (1931–) and Kent Wilcox in 1970, while in 1971 the first attempt to write out or sequence base pairs started with the lambda virus. Since viruses need few instructions to live, they were the logical choice for early sequencing work.

By 1972 the techniques of cellular control were well enough understood that scientists could begin manipulating them. Stanley Cohen (1922–2020), Herbert Boyer (1936–), and Robert Helling (1936–2006) placed foreign DNA in a host organism (in this case the bacterium E. coli), which then replicated itself. This was the technique of recombinant DNA, and it was the foundation of DNA cloning. In 1978 somatostatin became the first human hormone produced by recombinant DNA technology.

When Fred Sanger (1918–2013) introduced his chain termination method for DNA sequencing technique in 1977, essentially all the theoretical elements were in place for the big project: mapping the human genome. This idea had been around since the discovery of the structure of DNA, but the sheer size of the undertaking made it prohibitively difficult, and the tools to record or map the bases accurately

DNA

Bases Open

mRNA
Messenger RNA copies bases

Molecule being joined to chain.

tRNA
Transfer RNA assembles proteins

12.5 PROTEIN ASSEMBLY

were not available. As interest in decoding grew, new tools and methods were developed, and by 1983 a number of labs began to decode chromosomes in a variety of organisms. At about this time the American Department of Energy (DOE) was considering what to do with its various biological researchers. These scientists, many of whom were in the Office of Health and Environmental Research (OHER), had worked on those aspects of the nuclear weapons and nuclear energy programs that looked at the biological effects of radiation as well as doing basic research on cellular biology. These were not as pressing concerns by 1983 as they had been earlier, and so the DOE was looking for ways to employ them on other projects.

In 1985 Chancellor Robert Sinsheimer held a meeting at the University of California, Santa Cruz, to discuss the possibility of mapping the whole DNA list of bases in human chromosomes. The sequencing of the human genome would be a massive project, since there were an estimated 3 billion bases, and the average cost to sequence a base was about $10. Yet the potential benefits were enormous, from cures for disease to extended life. With the firm belief that techniques and equipment would improve, thus dropping the cost of sequencing, participants were enthusiastic about the future of the project. That same year OHER held a meeting at Santa Fe, New Mexico, on the feasibility of the Human Genome Initiative, while James Watson held a similar meeting at the Cold Harbor research center. Their conclusion was that the mapping of the entire human genome was both possible and desirable. It would be the Manhattan Project of biology, with big teams, big money, leading scientists, and a potentially world-changing result. At last, biologists could compete with physicists for money and prestige.

The following year, the DOE allocated $5.3 million for a pilot project. However, the National Institutes of Health (NIH) also began to fund genetics research. Given the medical and biological aspects of the research, it seemed strange to some officials that the DOE was running the show, but it had several advantages over the NIH. It had operated many large multilaboratory research projects so had in place a management infrastructure for big projects. It also had a great deal of money and political clout. Most of the conflicting issues were resolved when the DOE and the NIH signed an agreement in 1988 to work together on the project, which became international when HUGO (the Human Genome Organization) was formed to coordinate international research.

The year 1988 was the real start of the Human Genome Project (HGP), and it began with one of the discoverers of the structure of DNA at its head: James Watson, who held the position until 1992. Under his leadership, the HGP made great strides. Better computer technology, both hardware and software, was developed, and computer-automated equipment was integrated into research. The project was not only concerned with finding the code. It also addressed ethical and legal issues by allocating 3 per cent of grant money to fund research on the social implications of genetic research. In 1990 DOE submitted its proposal to Congress. Entitled *Understanding Our Genetic Inheritance: The US Human Genome Project*, it laid out a five-year budget of $200 million a year as the first phase of a 15-year project.

Because a significant portion of the budget was dedicated to research and the development of technology, the speed at which the sequencing took place was remarkable. The cost of analyzing bases dropped from $10 each to 10¢ each. The

HGP met its five-year goals in four years and by 1995 had high resolution maps of chromosomes 16 and 19, as well as large sections of 3, 11, 12, and 22. The following year a large international conference on the HGP, sponsored by the Wellcome Trust in Britain, one of the world's largest private funders of medical research, brought even more attention to the project. Global concern about the potential ethical issues associated with the HGP were serious enough that in 1997 UNESCO released the *Universal Declaration on the Human Genome and Human Rights*, which attempted to provide international agreement on the ethical use of genetic information.

In 1999 the HGP passed the 1 billion base mark, and with new advances in robotic sequencing equipment, on February 12, 2001, the HGP announced that it had completed what it called the "working draft" of the entire human genome. The genome sequence was completed in 2003, effectively ending the mapping part of the project, but many years of analysis and research remain to be done. This is only a general map based on a single individual, not a genetic blue print for all humanity. While the map is hugely important, the lasting legacy of the HGP will be the techniques and technology that were created to do this international project.

There were fewer genes, the functional parts of DNA that actually controlled cellular activity, than some scientists had anticipated, but they had decoded the owner's manual to the human organism. With recombinant technology and cloning, scientists could also turn bacteria and higher organisms into biological factories to produce drugs or other useful products. However, to understand why the development of recombinant genetics was more than an important scientific development, the laboratory work must be read along with two significant legal decisions: *Diamond v. Chakrabarty* and *John Moore v. The Regents of the University of California*. These two cases are not the only legal events concerning genetic material, but they illustrate why a kind of genetics gold rush developed.

On June 16, 1980, the Supreme Court of the United States decided in favor of the respondent in the case of *Diamond v. Chakrabarty*. It ruled that the Patents and Trademarks Office (USPTO), represented by Commissioner Diamond, was incorrect in its decision to deny a patent to Ananda Chakrabarty, a microbiologist working for General Electric. What made the case so important was that Chakrabarty's patent was for a microbe that he had genetically modified so that it "ate" parts of crude oil, converting them into harmless by-products in order to clean up oil spills. In effect, the Supreme Court ruled that the microbe was not naturally occurring and was a new and useful "composition of matter." The decision was a close one, going five to four in favor of the patent. Although the Court offered a fairly narrow window for the patenting of living organisms, it nonetheless granted the legality of modern genetic patents, and the race was on for genetic commerce. The decision overturned the long-standing policy of the USPTO of denying patents on living things.[1] Since the USPTO had become the *de facto* venue of record for patents, a policy change in the American patent rules had far-reaching implications for everyone. International patent agreements recognized the protection of patents granted in one country as being valid in all signatory countries.

The USPTO continued to deny patents on living things until 1987 but then issued a formal policy statement allowing the patenting of nonhuman multicellular living organisms. This was partly in response to the Chakrabarty decision and other cases, but it was also a change to promote genetic research. Many researchers, universities, and private companies had argued that without patent protection there would be less incentive to do genetic research and other countries might gain an advantage in the area.

Among the first "higher" organisms to be patented was the Harvard mouse, also known as Oncomouse. This mouse was genetically modified to be more susceptible to cancer, and it was sold as a tool for cancer research. Its creators were granted an American patent in 1988.[2]

The second important case was *John Moore v. The Regents of the University of California*. John Moore, who suffered from a particular type of leukemia, was treated at the UCLA Medical Center in 1980 and had his spleen removed. He was also asked to return several times for tests and to give other kinds of tissue samples. Moore's doctors felt that his tissues or cell lines might be extremely useful for research, so they developed research material from his spleen and the other samples. In 1981 they applied for and were granted a patent on his cell line, which was then sold commercially. Three years later Moore discovered that his cells were patented and being sold without his knowledge, consent, or financial return, so he sued. In 1990 the Supreme Court of California ruled on the case and effectively denied the suit.

This ruling declared that in California, once cells were out of a person's body, they no longer belonged to the person they came from and that anyone who got them could claim them as their own property. To many, the ruling seemed absurd on several levels. First, if a person lost a hand (a collection of cells) in some terrible accident and someone came along and took the hand away, that would clearly be theft; yet, in *Moore v. Regents*, the court ruled that researchers taking cells were not doing anything illegal. Second, the UCLA lawyers argued that it was

---

1. The USPTO police was not without exception, since a patent had been granted to Louis Pasteur for a type of yeast in 1873.
2. In 2002 the Supreme Court of Canada denied an application for a patent on the "Harvard mouse," arguing that higher life forms could not be considered new inventions.

in the public interest to allow genetic material to be collected and used by researchers but that it would be against the public interest if those researchers had to keep track of and compensate the original source of the material. The added cost and work would inhibit research. Thus, their argument seemed to be that it was good for the public interest that universities and private companies profit from genetic research, but it was bad for the public interest that the public (the source of the genetic material) profit from genetic research.

A flashpoint for many of the concerns about genetics was the decision in 1998 by the Icelandic government to sell, or perhaps more correctly lease, the entire genetic heritage of the Icelandic people to a private company. Kari Stefansson, a genetic researcher, established the company deCODE and, in part funded by $200 million from the pharmaceutical company Roche, proposed the creation of the Iceland Health Sector Database (IHD). This database would include genealogical records going back hundreds of years and public health records starting in 1915, as well as genetic information on almost all Icelanders. Iceland made a good target of study for several reasons. The population is relatively small, around 275,000 people, and relatively homogeneous. Although a tiny country, it is a First World nation with universal literacy and a high level of general education and public health care, as well as the longest parliamentary tradition in the world. The extensive genealogical records that reach back to the Norse era allow researchers to identify genetic groups historically and to correlate them with genetic information from the current population.

Supporters of the project argued that Iceland benefits from deCODE's activities by getting money, high-technology research facilities, genetic information, and free access to any drugs or therapies that might arise from the information in the IHD. Detractors argue that the government has overstepped its authority and thus violated human rights by collecting the information (and in fact outlawing the private sale of genetic information by individual citizens) and that there are problems with the privacy system used to protect patient confidentiality, since DNA is by definition the perfect identifier. The government has allocated to itself the right to gather and use genetic information as if it were no different than information for a driver's license, census data, or tax returns.

As ambitious as Stefansson's project was, it lacked a clear business model. In 2012 deCODE was in financial trouble having never turned a profit and was purchased by Amgen, a biopharmaceutical company based in California. The genetics system and database part of deCODE was then sold to WuXi PharmaTech in 2013. deCODE Genetics of Iceland continues as a subsidiary company of Amgen, continuing the search for genetic markers for disease.

In a wider sense, some commentators have pointed out that what the IHD provides is not just a tool for the detection of disease but a baseline for what is a "normal" human. The healthy, blond-haired, blue-eyed Icelanders strike some critics as a potential model for human appearance. This fear, played out in any number of science fiction horror stories such as the film *Gattaca* (1997), has in turn been dismissed by supporters of genetic research as both unlikely from a technical point of view and socially undesirable. The social implications of genetics, whether modified canola or human beings, offer a frightening vision of the future for some and a potentially utopian future through the mastery of nature to others.

Conclusion

Whether it is the power of computers or the potential of the genetic revolution, the utility of science had been brought home in the most direct way by the end of the twentieth century. Science and society are now inexorably intertwined. The public has come to expect that science will produce the consumer goods being sought by an increasingly affluent society, and in the creation of many of those goods, such as the transistor and the computer, new tools for research and avenues of inquiry have been opened. Scientific investigation of the natural world helped environmentalists think about the planet as a closed ecosystem and argue for better stewardship of the environment. At the same time, science has developed new ideas and techniques for manipulating that very nature we are supposed to be preserving. For good or ill, we now live in a world where the scientific point of view takes precedence over most other ways of knowing the world. Science has, in a very real way, made the world we live in and changed it forever.

Essay Questions

1. Was Ada Byron, Countess of Lovelace, the first computer programmer?

2. How did solid state physics transform modern consumer electronics?

3. Why did some biologists think that the HGP would be like the Manhattan Project for biology?

4. Why is Rachel Carson's book *Silent Spring* seen as one of the founding documents of modern ecology?

CHAPTER TIMELINE

1895 ○ Konstantin Tsiolkovsky suggests cable to space – first conception of a space elevator

1956 ○ human growth hormone isolated

1966 ○ First scientific discussion of a space elevator

1972 ○ Recombinant DNA

1981 ○ Scanning tunnelling microscope – can manipulate individual atoms

1983 ○ Huntington's disease identified as a genetic anomaly

1985 ○ Discovery of buckminsterfullerenes

1989 ○ "Cold Fusion" controversy

1990 ○ Genetically modified T cells used to treat adenosine deaminase deficiency

1997 ○ Kyoto Protocol on climate change

1998 ○ Dr. Nancy Olivieri and drug deferiprone controversy
Andrew Wakefield makes anti-vaccine claim

2007 ○ IPCC wins Nobel Prize

SCIENCE AND NEW FRONTIERS: POTENTIAL AND PERIL IN THE NEW MILLENNIUM

13

I n 1995 the Natural Law Party ran candidate John Hagelin, a Transcendental Meditation disciple with a doctorate in physics from Harvard University, in the American presidential election. During the campaign the party offered a scientific demonstration of the power of Transcendental Meditation to bring peace to a troubled world. More than 5,000 disciples gathered in Washington, DC, to meditate, with the purpose of bringing peace to a notoriously violent city. A year later the party released a study that "scientifically" proved that Washington had indeed been more peaceful because of the waves of love and harmony projected by the meditators. When it was pointed out that the crime rate in the city during the demonstration was actually much higher than normal, the party defended its study on the basis that, without the meditators, the crime rate would have been even higher.

The case of the yogic peace waves highlights the power and problem of science at the beginning of the twenty-first century. Because we have come to accept and even expect wondrous things to come from science, it has become easier to make wild or even fraudulent claims of discovery so long as the claim is draped in scientific terms or supported by someone who claims to have scientific credentials. If a person with a PhD in physics says something is scientifically proven, shouldn't the general public accept the reality of the claim? After all, what higher standard of scientific expertise is there than a doctoral degree in physics?

Our understanding of what science is and how it interacts with our society has been fundamentally altered. For the Greeks, the study of nature was an elite,

highly controlled activity, undertaken by a tiny intellectual cadre with philosophical and religious objectives. For people living in the twenty-first century, science is
a public activity that has become an incredibly powerful tool for political, economic, and social change. The success of science has been so great that people
seek to link themselves with it even when there is little or no actual science being
used. When the Natural Law Party or the makers of cosmetics claim the mantle of
science they are hoping that the association will lend an air of truth and reliability
to what they are offering. This "promotion by association" clouds the public
understanding of science, often to the point where the word "science" is reduced
to an advertising tag similar to "fresh" or "wholesome."

How did this change take place? Gradually, state leaders – princes and
governments – began to see the utility of science and scientists, either to elevate
their status as cultural and intellectual leaders or to boost their military and
economic power and well-being. Sixteenth-century natural philosophers, providing status and spectacle for European princely courts, are the linear ancestors of
the Manhattan Project and Big Science funded by the government and military
today. Scientists themselves contributed to this transformation, arguing for the
usefulness of their investigations even when this was not the case. Here, the
parallel between the rhetoric of early Royal Society statements and that of modern
grant applications is instructive. Granting agencies want to know what they will
get out of the relationship in exchange for the money and connections they offer.
As the power of scientists has increased, so has the value of association with
science.

Yet the very success of science has created a backlash. Although there have
always been people who reject science on religious grounds, today there is a
growing number who reject science for political or more broadly, ideological
reasons. Some science deniers reject specific scientific ideas such as smoking as a
cause of cancer or climate change, while others reject the very premise of scientific knowledge. For example, supporters of the idea of a flat earth or the
anti-vaccination movement reject the concept of objective knowledge. For them,
personal feelings are superior to any body of tested data. In addition to the
rejection of the real world, there is often an associated set of conspiracy theories in
which doctors, scientists, governments, and the "mainstream" media have some
secret agreement to keep the public ignorant.

The result of this transformation of science is thus paradoxical. On the one
hand, twenty-first-century science has the power to transform our lives and our
understanding of the universe in ways we can barely imagine. The huge resources

now available to scientists, the tools at their disposal, and society's belief in the beneficence of science in general all contribute to the potential flowering of fascinating and significant scientific results. On the other hand, the success of science has led to a credulity that allows scientific charlatans to flog their wares with impunity and to a growing number of people whose fear of science is so great they reject science as a way of producing reliable knowledge.

The power of science, the legitimate concern about its misuse, and a gullible misunderstanding of scientific principles are all the result of the triumph of science since the scientific revolution. The challenge now will be to expand our understanding of the idea of science and to use our informed judgment to improve the world and keep it safe.

What Is Science in the Twenty-First Century?

One of the ways we can see the transformation of science in our society is in the act of talking about science, which has become more complicated. The term "science" has been remade. While it has never had a universally accepted definition, it once was commonly regarded as referring only to the study of the physical world and the tools and methods of that study. The term has now become a general indicator of a claim for any profound or specialized knowledge. Adding "science" to a phrase is often an attempt to make whatever it is attached to seem more certain, insightful, true, or useful. The phrases "science of hair care," "the science of business management," "political science," or "created by a leading scientist" attempt to tie a product to the idea of science, even if under minimal scrutiny there is no practical or historical link to science whatsoever.

Even before the members of the Royal Society and the Académie des Sciences began to promote the concept of the utility of science as a justification for the existence of their organizations, the utility of science had been integral to the ethos of research. Patrons were looking for more than philosophical insight when they employed natural philosophers, and, like Grand Duke Cosimo, we have come to expect science to provide more than esoteric knowledge. The successful exploitation of science has been demonstrated so powerfully over the last 400 years that nations now neglect it at their peril. It has changed the course of wars, helped raise up countries economically, and changed gender relations. Science is now so closely linked to the success of nations that it is a mandatory subject of education in all industrialized nations and many others. Every child must learn science to gain

CONNECTIONS

Indigenous
Worldviews:
Religion and
Science

Over millennia, Indigenous people around the world have developed rich and complex understandings of the world around them. As western science gained authority and power, western thinkers categorized these worldviews as spiritual rather than as scientific. This did not stop western researchers from appropriating the natural knowledge of Indigenous people and using it to aid colonial expansion and add to scientific knowledge. Both the simplistic categorization and the appropriation of knowledge have led to a complete misunderstanding of the complexity of Indigenous worldviews. In these worldviews, there is a fundamental, rich, and necessary interaction between a spiritual understanding of the world and an understanding of its natural functions. Likewise, this dismissal of Indigenous knowledge ignores the fact that such knowledge can work in partnership with Western scientific knowledge to generate new understandings of the natural world.

Many North American Indigenous religious and spiritual practices are based on an understanding of a continuous and continuing act of creation, in which humans play a role in the cycle of creation. This creation is not one act, but rather it is at the same time a power, an action, and an energy involving everything in the universe. This understanding implies that the world is in a constant state of coming into being, a way of understanding cosmology that

productive employment and to be a good citizen. The degree of integration of science education is one indicator of what separates the developed from the developing world.

Science has been so broadly injected into industrial society that it is difficult to distinguish what science is and even who is a "scientist." While we readily acknowledge the science Nobel Prize winners as scientists, does the category include people with advanced degrees who do no original research, such as purity control chemists or physicists who work on Wall Street? Physicians receive a great deal of scientific training, so are general practitioners scientists or does the term apply only to medical researchers? Psychoanalysts, homeopaths, computer programmers, and sociologists have at one time or another claimed scientist might seem similar to some modern discussions of early universe creation in Western cosmology.

Likewise, Indigenous concepts of nature and creation are often place specific; knowledge and enlightenment come to particular people in particular spaces and these spaces then become sacred and powerful throughout the creative cycle. Modern Western religions and their related concepts of nature, on the other hand, have tended to reduce nature to universal truths, eternally existing outside the particularities of space or time. These views of nature, and human participation in nature, have fundamental implications for the interpretation of nature and its laws.

Many Indigenous people have engaged in deep and close observation of the natural world over many years. The specificity of their observations are of first importance to them, while Western scientists have tended to look for universals. It is only by bringing these two ways of knowing into conversation that the real experience of ecological change, for example, can begin to be understood. When Indigenous people observe that fish biology and behavior have changed over time when downstream from a mine, this is the result of prolonged and careful experience of these species in particular places. Modern scientists have tended to approach such problems with sampling and statistics, but are now beginning to acknowledge the importance of local observations. A "fly in, fly out" way of doing field research is less satisfactory than research by people on the land who see themselves as connected to the things being studied.

As governments in settler communities such as Canada and Australia seek to find paths to reconciliation with Indigenous populations, more conversations about alternate worldviews, ways of knowing, and knowledge itself will need to be facilitated. If we think of this as the creation of a "trading zone" of knowledge, where both sides can come as equal partners, to truly exchange rather than expropriate or belittle, this may lead both to better appreciation for different ways of knowing and to the expansion of natural knowledge going forward.

status. A hominid paleobiologist and a cultural anthropologist may work in the same department at a university, but are they both scientists? Clearly, there is a spectrum of careers from "very scientific" to "no science required," but the range and number of people claiming science status has expanded enormously.

The functional definition that "science is what scientists do" has started to break down in the modern world and will break down even more in the future as the trend to claim scientific status by a wider and wider range of people continues. This will make the problem of informed choices about scientific issues even harder. In the era of the greatest number of working scientists and the widest teaching of science in human history, the very concept of scientific knowledge can become muddled. The number of people claiming to offer scientific insight has become legion, so there is frequently conflicting "expert" opinion on a variety of socially important issues. Whether it is in a murder trial or a debate about global warming, scientific experts with equivalent credentials may offer diametrically opposed opinions. As science has had a greater direct effect on society and more people can claim to be doing science, we have seen a rise in the misunderstanding, or misrepresentation, of scientific ideas. Further, outright fraud cloaked as science has become easier. Any gap in scientific knowledge can then be held up as a failure of science and used as a rationale for rejecting science.

Science Must Perform Miracles: Genetic Testing and Nanotechnology

Governments, industries, and ordinary people all believe that science should do wondrous things. Some of the miracles are big, such as the discovery of insulin to treat diabetes, saving millions of lives, while other breakthroughs are small and often go unnoticed by the general public, such as the discovery of a new species of ichthyosaur that resembles a cross between a crocodile and a dolphin. Two areas that promise major discoveries that have a direct effect on human life are genetic testing and therapy, and the materials revolution in nanotechnology. While each offers the potential for great benefits, each also carries with it concerns about application, especially ethical issues about transforming the human condition.

Genetic testing is, in some ways, a continuation of other forms of testing for diseases. Whether it is culturing blood to discover the presence of a disease organism or looking at tissue samples for signs of cancer, the first stage of genetic testing was based on looking for an indicator of some problem. For example, in 1983 James Gusella (1952–) and his team identified Huntington's disease as a genetic anomaly on chromosome 4. As mapping has improved, the number of problem sites that can be identified has increased dramatically. Hundreds of tests now exist for diseases such as cystic fibrosis, Tay-Sachs, Down syndrome, and many others. Tests are now so precise and therapies so specific that they have opened the door to a new form or medicine called "personalized medicine" or "precision medicine." With test results and genetic information treatments unique to the patient can be created. Rather than the "one shot for all" treatment of early antibiotics, custom-made medicine and gene hacks designed for a specific patient are becoming possible and practical.

Testing is also starting to reveal not only existing conditions but also the potential for disease. This means that a person might, for example, have an elevated risk of getting a particular kind of cancer, although it is not a certainty. This raises ethical questions, since genetic testing could be used to formulate public health policy or on a more individual level be used by health insurance companies to determine the cost of coverage based on the likelihood of a person being affected by genetically linked diseases.

There is also an ethical issue regarding the ownership of tests. Since genes can be patented, tests based on certain gene lines have become private property. Patent holders have forced laboratories, both medical testing labs and research labs, to pay fees for certain tests or stop doing them, not because equipment or even techniques are being used without compensation, but because of ownership of specific genetic material. Who owns the genetic material of the globe is being hotly debated, especially in cases where the genetic material has been gathered from Indigenous people, who may or may not have been informed of its potential use.

Genetic testing of a wide population has also been made possible, raising issues of privacy versus social benefit and of who has the right to information. Health care providers might be able to plan more rationally for services if they had genetic information about the general population, but such broad testing might also be used to deny insurance to people likely to develop costly health problems. Should employers be allowed to test employees to see if they are likely to get sick? Should potential parents be screened before they have children? Equally, a number of researchers have claimed that there are genetic links to behavior, so genetic testing might be used to identify people who are more susceptible to addiction or criminal acts. While it may sound like a bad science fiction story, we have the technology to create mass genetic databases on whole national populations, as was done in Iceland (see Chapter 12) and which is limited only by computer storage capacities. In addition, genetic mapping has become so inexpensive that companies are now offering test kits for sale to the general public that purport to tell you your heritage and (where allowed) potential health issues. These commercial tests pose another version of the question of who owns genetic information, since the testing companies retain and aggregate test results and market that information. In this case, the person who buys one of these tests, often as a gift for someone else, is not really the consumer, but the product being marketed to others.

While testing can indicate problems, it is the ability to use genetics to repair problems that now attracts the most interest. The first use of recombinant DNA in therapy was undertaken by a team working for the NIH. In 1990 and 1991 they used retroviruses to modify T cells (part of the immune system) taken from two girls with a rare genetic disorder called adenosine deaminase deficiency. When the T cells were returned to the patients, a certain portion of the T cells being produced by the patients continued to be free of the genetic problem. Their health improved to the point where they were able to reduce their drug intake to half the amount used in conventional treatment of the disease. While this was not a complete cure, it was a remarkable change for the patients.

By 2014 more than 2,000 protocols for gene therapy had been conducted by the NIH. Many of the ethical concerns about genetic therapy have to do with limits to its use. While treating some genetically based diseases seems unproblematic, at what point is a therapy no longer about defense against disease but about changing a person in order to reach some desired condition? Should gene therapy be developed for baldness or to make people taller?

The case of height is an interesting one, since it bridges both the pre-genetic and genetic treatment eras. The hormone HGH or human growth hormone was isolated by Choh Hao Li (1913–87) and his team in 1956. While HGH is not the only control mechanism that affects human height, it was developed as a treatment for dwarfism in children in the 1960s. Over time doctors started getting requests for HGH treatment for children who did not suffer from dwarfism but were just shorter than average height or even at average height. This raised ethical questions about the use of therapy to improve the human body rather than ameliorate a debilitating condition.

The uncertainty associated with genetic therapy and problems such as the nonclinical use of HGH led the NIH to establish a panel to examine the issue. As the history of HGH shows, the ethical questions are not theoretical. Although the panel's 1995 report argued that genetic therapy had many potential benefits, and future work looked very promising, it also warned people about too much enthusiasm:

Overselling of the results of laboratory and clinical studies by investigators and their sponsors – be they academic, federal, or industrial – has led to the mistaken and widespread perception that gene therapy is further developed and more successful than it actually is. Such inaccurate portrayals threaten confidence in the integrity of the field and may ultimately hinder progress toward successful application of gene therapy to human disease.[1]

1. Stuart H. Orkin and Arno G. Motulsky, *Report and Recommendations of the Panel to Assess the NIH Investment in Research on Gene Therapy* (Bethesda, MD: December 7, 1995).

In 1985 the Food and Drug Administration approved the marketing of HGH produced by genetically modified bacteria, the second genetically engineered drug after insulin to be introduced. While geneticists and physicians worked on these products with therapy in mind, promoters of the use of HGH touted it as a wonder drug that would increase muscle mass, decrease body fat, make a person look and feel younger, and even increase libido. While these claims are largely untrue, or at best unsubstantiated, promoters have used the scientific foundation of HGH work to justify wild claims and to send e-mail spam advertising their product to millions of people. Marketing and a fear of failing to keep up genetically with the next-door neighbors may drive the demand for therapy and provide ample opportunities for charlatans offering a kind of elixir of life.

Genetic manipulation is no longer a theoretical concern. In 2018 biophysicist He Jiankui (1984–) announced the birth of twin girls whom he claimed to have used CRISPR gene editing on embryos to increase their HIV resistance. The announcement led to a storm of media interest and great concern about the ethics and scientific reliability of He's work. In 2019 a third child with edited genes was born. Working with a small team at the Southern University of Science and Technology, He's work was first hailed as a breakthrough, but it quickly became clear that there were problems. His university stated that his work was conducted outside of the university, and the project was condemned by leading scientists including Nobel laureate David Baltimore (1938–), a pioneer in genetic research. In 2019 a Chinese court found He and two of his associates guilty of ethical violations and misleading the medical doctors involved in implanting the gene-edited embryos. Despite He's claim of success, subsequent testing of the children raised questions about the degree of cellular change suggesting little or no protection from the editing. There has also been concern about the possibility that the changes had affected the brains of the children in unpredictable ways.

As our knowledge of cell function increases, it is likely that genetic therapies will improve. One of the principal targets will be cancer cells, where the ultimate goal will be to get the cancerous cells to return in effect to their normal state. There is, however, another path to dealing with the problems of cells, and that is to create them to order rather than trying to fix them up later. Genetic modification, which has already been introduced to a number of food crops and the production of the Harvard mouse, as well as drug-producing bacteria, will come to humans. Some observers, such as Jeremy Rifkin (1945–), have already speculated on the potential smorgasbord of genetic choices that may be available to parents in the future. Everything from eye and hair color, disease resistance, breast size, height, length of natural life, intelligence, and even musical ability may be modifiable. While the movies have tended to portray the genetic modification of people as an evil plot to create a super race (often in the form of relentless killing machine–soldiers), in reality the choice to modify human fetuses will be based largely on parental concern about providing the best life possible for their children. What will the world look like when the rich can modify themselves and the poor cannot?

The Materials Revolution

In 1959 the physicist Richard Feynman (1918–88) gave a lecture entitled "There's Plenty of Room at the Bottom," about the possibility of the direct manipulation of atoms as a way of doing synthetic chemistry. This has been seen as the start of nanotechnology both theoretically and in practical terms because Feynman concluded his talk with two challenges: (1) Make a working electrical motor that would fit in a cube with 0.39 mm sides and (2) Reduce a page of text to be 25,000 times smaller than normal print. The first challenge was finished in 1960 by William McLellan (1924–2011), but it took until 1985 for Tom Newman to inscribe the first page of *A Tale of Two Cities* on the head of a pin using an electron beam. At that size, the whole of the *Encyclopedia Britannica* could be written on the pin.

The main material that has become the focus of nanotechnology has been carbon. Carbon as a structural material really started with the introduction of carbonized polyacrylonitrile by Akio Shindo (1926–) in the early 1960s. By bonding carbon fibers together and making thread and cloth from carbon, it could be used to replace heavier materials. With a tensile strength equal to steel, but only a fraction of the weight, carbon fibers have increasingly become the material of choice for designers and builders. Although there were some problems with delamination and brittleness in the early carbon fiber components, by the 1990s better polymers were being used and carbon fiber was being used in everything from tennis racquets to airplane wings.

While carbon fibers are part of nanotechnology, it was another form of carbon that started a new carbon revolution and was first called nanotechnology. In 1985 Harold Kroto (1939–2016) at the University of Sussex was wondering about carbon chains in space. Evidence suggested that certain stars, red dwarfs, produce a kind of soot. If this were the case, these carbon chains would be one of the oldest possible molecules, perhaps forming the foundation for a number of celestial objects and providing the materials to make up organic matter in the universe. To test this hypothesis, Kroto asked Richard Smalley (1943–2005) and his team at Rice University in Houston to recreate some of the conditions that exist at the surface of a red dwarf star. By shooting a laser at a block of carbon and collecting the vaporized clusters of molecules, they discovered that some of the molecules contained a fixed number of carbon atoms, either 60 or 70. The resulting molecules looked like geodesic domes as designed by R. Buckminster Fuller (1895–1983) and were formally named *buckminsterfullerene*. (See figure 13.1.) Fullerenes or more simply, bucky balls, have several properties that make them interesting. They can conduct electricity, are very hard, and because of their shape, can capture other atoms.

13.1 BUCKMINSTERFULLERENE OR "BUCKY BALL"

Perhaps the most audacious use suggested for such carbon fibers was to create a cable 99,820 kilometers long rising up from the equator to a space platform. The idea of such a system was first visualized by Konstantin Tsiolkovsky (of Russian rocketry fame) in 1895 when he looked at the Eiffel Tower and imagined a cable rising up from it into space. One of the first examinations of the possibility of a space elevator came from the team of John D. Isaacs (1913–80), Allyn C. Vine (1914–94), Hugh Bradner (1915–2008), and George E. Bachus (1930–), who published "Satellite Elongation into a True 'Sky-Hook,'" in the journal *Science* in 1966. (See figure 13.2.) In addition to moving materials cheaply, the space elevator would also serve as a launch platform, flinging things into space at a velocity of more than 25,000 kilometers per hour, like a giant slingshot using the rotation of the Earth.

Although nano fibers are important, the ultimate goal is to create nano devices. A start toward this submicroscopic engineering was made in 1981 when Gerd Binning (1947–) and Heinrich Rohrer (1933–2013), working at the IBM Research Lab in Zurich, created the first scanning tunneling microscope (STM). Although it is called a microscope, it examines materials well below the range of any optical system. Rather than looking at small objects, it feels them in a way analogous to a phonograph needle feeling the contours in the groove of a vinyl record. With a current running through a very sharp needle, the STM can trace the outline of objects down to a hundredth of a nanometer (a millionth of a millimeter). High-powered computers turn the data into a visual image.

13.2 SPACE ELEVATOR

This microscope has a second trick up its sleeve, however. In addition to sensing small objects, it can pick them up and move them around. In 1989 another IBM team at Almaden, California, used an STM to arrange 35 xenon atoms to spell "IBM." This feat was topped in 1996 back in Zurich when scientists made a miniature abacus consisting of 11 rows of ten C_{60} fullerene molecules, which the tip of the STM pushed backwards and forward to count. While an abacus that requires an STM to count may seem like a bit of a science stunt, it actually has a serious purpose. If nanotechnology is to be created, there has to be some way to make the original nanomachines. In a sense the STM offers the possibility of creating the machine shop that will produce the machine-making machines. Other researchers have developed microscopic pumps and motors, so the production of such micromachines seems entirely possible. Some scientists foresee nanofactories creating useful materials from piles of raw chemicals, from the molecular-level assembly of carbon-fiber materials to whole objects – chairs, computers, and airplanes – out of vats of basic elements.

A kind of macro version of this has already appeared with the invention of additive manufacturing or 3D printing. In 1981 Hideo Kodama (1944–) of the Nagoya Municipal Industrial Research Institute introduced a method of making three-dimensional objects using a liquid plastic that would harden on exposure to ultraviolet light. Since then a variety of printing methods have developed including sintering (using heat to turn a powder to a solid, often using metal), liquid films, and thermoplastics. While the majority of additive manufacturing is still directed at making prototypes or specialized components, consumer machines are becoming available. This technology has the potential to be used for nanoscale production, replacing the extremely expensive STM approach with more commercially affordable production methods.

Good Science Goes Bad: Cold Fusion

There is a dictum in both science and law that extraordinary claims require extraordinary evidence. In the case of cold fusion, the claims were indeed extraordinary while the evidence was not. The pressure to get ideas out of the lab and into the market has opened the door to scientists taking shortcuts to preserve priority and commercialize discoveries. One of the most notorious cases highlighting the dangers of circumventing established scientific processes is the story of cold fusion.

Given our general expectation that science can produce wonders, it did not seem completely unrealistic in 1989 that a new energy source had been discovered. When Chase Peterson, president of the University of Utah, announced that two scientists, Stanley Pons (1943–) and Martin Fleischmann (1927–2012), had discovered fusion at room temperature, it caused a sensation. They had created an electrolysis "cell" consisting of a glass flask containing heavy water (deuterium oxide) and run a current through it. At some point, they noted a sharp rise in temperature even though the amount of power being used had not changed. If the scientists had generated even a tiny bit more energy in their fusion cells than they had put in, their accomplishment could revolutionize energy production, rewrite physics, and make large sums of money. They called their discovery "cold fusion" as opposed to the "hot fusion" seen in stars or hydrogen bombs.

In the days immediately after the announcement, it was not clear what Pons and Fleischmann had actually done or even the exact experimental setup they had used, and so the idea that some new aspect of physics had been uncovered seemed possible. Some hastily constructed experiments at other labs appeared to confirm, or at least not clearly disprove, the claim. If cold fusion was real, some parts of physics would have to be revised, but such events had happened in the past. The unexpected results of a simple experiment heating up a black cube of carbon had, for example, helped initiate quantum physics. While most of these kinds of scientific controversies take place within the scientific community, Pons and Fleischmann played out their story in the full glare of media attention and with the backing of tremendous amounts of research money. Part of the controversy came from other scientists, especially physicists, who said that cold fusion could not possibly work, setting up a fight between the chemists (Pons and Fleischmann) and the physicists (such as John R. Huizenga [1921–2014]), which provided an easy conflict to cover in the media.

The stakes of scientific discovery are high: not only money, equipment, and research time are being used, but people's reputations are at stake. It was thus important for Pons and Fleischmann to demonstrate publicly their precedence in this revolutionary scientific discovery, without having gone through the time-consuming process of checks and balances required by peer review. In addition, patrons, in the form of governments, philanthropic organizations or private companies, must make decisions about what types of research they will fund and which people within those areas should be supported. Those patrons are keen to have publicity, demonstrating that their funding has been valuable in the creation of new and potentially useful knowledge. Although funding is not quite a zero-sum game in which the total amount of money available is fixed, it is close to that, so backing the wrong research project or the wrong people can damage the potential for future funding. While it is difficult to calculate how much money and time has been spent on the chimera of cold fusion, it is in the tens of millions of dollars and the loss of thousands of laboratory hours that could have supported other research efforts.

Claims for cold fusion turned out to be completely unfounded, and there was no challenge to our understanding of physics, but the event highlights some of the problems of science in the new millennium. The major characters in the story, Pons and Fleischmann, were reputable scientists with degrees, memberships in professional associations, and employment at major universities. They had all the credentials to make them reliable sources for scientific discovery, so it was not unreasonable to presume that their work merited serious consideration. That other scientists objected to their conclusions did not, in itself, make the work wrong. History is littered with stories of established scientists opposing new discoveries, from Priestley's objection to Lavoisier's oxygen theory to the controversy over Einstein's theory of relativity, or the geological community's condemnation of Wegener's idea that continents move. In their eagerness to establish precedence for a potentially revolutionary idea, however, Pons and Fleischman believed that they needed to present their ideas as proven before they had been subjected to rigorous investigation by others. This demonstrates the danger of a system that has increasingly commercialized research in science by offering great rewards to those who can produce extraordinary results quickly.

As a postscript to the cold fusion story, a small number of private and public groups are still funding cold fusion research despite more than 25 years of failure. This is often held up by supporters as proof that there is a scientific foundation to the original idea, but in fact such funding is a form of marginal investment. In other words, patrons fund a few unlikely projects as a bet on a long shot, particularly in tax jurisdictions that allow them to write off research costs.

Science in the Corporate World

Most commentators have attributed the "discovery" of cold fusion to wishful thinking and poor experimental procedures rather than malfeasance. Science does have a method of self-regulation, designed to squeeze unsound science out of the range of topics considered by scientists as legitimate or worthwhile. Scientists point to the peer review system in journals (not used by Pons and Fleischmann, who turned instead to the mass media) and to the replication of experiments as the internal means of weeding out bad science. This is, however, more problematic than you might think, since it turns out to be difficult to repeat big experiments, and few experiments are ever repeated in practice. Since scientists' livelihood is based on discovery, there is not much support for repeating work that has already been done. As well, since the cost of large-scale experiments, such as supercollider tests, can run into the millions of dollars, it may not be possible economically to repeat certain experiments even if there were some interest in doing so. Scientists, therefore, often rely on the consilience of experimental results, rather than on repeated experiments. In other words, experimental results are accepted as valid, even if not independently tested, as long as the results fall within expected norms, corresponding to already established theories, and the experiment has been conducted following accepted procedures.

A second and growing problem for self-regulation is the use of secrecy in research. While aspects of research of military importance have long been kept secret, more nonmilitary research is being kept confidential, based on the idea of proprietary information not only in corporate-funded research but also increasingly in public research, as universities and governments look to spin off research into profit-making businesses. If there is no evidence to evaluate, claims are hard to test, and timely decisions about research or products become difficult. Secrecy can affect both the discoveries, which are often shielded from public scrutiny, and the application of research, also kept secret. Legal settlements over product liability that have used scientific research often have included nondisclosure agreements or "gag orders" to prevent parties from revealing to others what problems were uncovered. Such restrictions have been applied to a range of products, from cigarettes to cosmetics. If scientific work cannot be examined by other scientists, then the self-regulation system breaks down.

Added to this problem are efforts by interested parties to protect their investments by interfering with scientific inquiry. This can take the form of biased research or efforts to block research that might indicate problems with a product or procedure. The most publicized cases have been in the pharmaceutical business, in which a number of researchers have falsified results in favor of a drug. Conversely, other researchers have been fired, threatened with legal action, sued, or had funding withdrawn for publishing negative results or suggesting problems with drugs. Such was the case for Dr. Nancy Olivieri, who was threatened with legal action and removed from her position at Toronto's Hospital for Sick Children after publishing a negative report on the drug deferiprone in the *New England Journal of Medicine* in 1998. Although she was later reinstated, her case was far from isolated.

In the anti-vaccine scandal starting in 1998, surgeon and medical researcher Andrew Wakefield (1957–) claimed at a press conference to have found a link between the measles, mumps, and rubella vaccine (MMR) and autism. His research was published in the journal *Lancet*, but was later discovered to have been poor quality and misreported to support his position, leading to the withdrawal of the paper by the journal. It was further discovered that he had undisclosed financial interests, receiving money from lawyers in a lawsuit against the MMR vaccine and connections with two pharmaceutical companies, one of which was planning to develop an alternative vaccine.

The problem of biased research has become so urgent that in 2001 the International Committee of Medical Journal Editors issued a warning that they would no longer publish drug trial reports from researchers who were bound by agreements that limited academic freedom. In other words, it was the whole story or no story.

Since new drugs can cost millions of dollars to produce and can generate billions of dollars in revenues, the pressure to publish positive research is very great. An increasing number of scientific journals now require a disclosure of financial interest, such as funding sources or corporate remuneration, from scientists submitting papers.

While interested parties may try to circumvent the publication of negative results, science journals are not always neutral players, so the self-regulatory mechanism of peer review is not completely reliable either. In recent years, even the most prestigious journals such as *Science* and *Nature* have been accused of rushing results into print in order to be the first to publish cutting-edge work. Because of the importance of publication, a kind of feedback loop exists between the scientists and the journals. Journals gain prestige by publishing exciting and groundbreaking results. Scientists who want to make a name for themselves aim to have their work published in prestige journals. This convergence of interests is not necessarily a problem, although it can encourage both players to take short-cuts. As well, the reliance on "blind" or anonymous reviewing for publication is not always impartial. In fields where everyone's work is known to everyone else, the control mechanism of peer review may break down. Since only scientists in the same field can understand and evaluate work, when they act as reviewers they may be reluctant to criticize other members of the same small community. The editors of scientific journals have little defense against bad work if the peer reviews on which they rely are biased or incomplete. The stakes are high. Millions of dollars in research money, leading research posts, international prestige, and the ultimate prize – a Nobel Prize – may ride on the status of publications.

The various problems of verification are a natural by-product of a complex and self-regulating profession, but it can make the interaction between scientists and the general public difficult. Laypeople are left troubled by conflicting and often contradictory scientific claims, so it is no wonder that some have rejected the whole enterprise of science, greeting all new discoveries with skepticism. Ignoring science is putting one's head in the sand as it continues to change our lives, whether we like it or not. In the coming years, we as individuals and as a society will be faced with an increasing number of choices presented to us by science. Where should research money go? How will we assess the differential importance of demands for cash for a giant synchrotron to look into the interior of subatomic particles against space stations or the search for the cure for cancer? How do we balance the dangers against the potential benefits of creating genetically modified food in a world where starvation and pestilence are common? How do we assess arguments about global warming? What about deeply personal choices, such as the potential to genetically modify our children or even ourselves?

Denialism

One of the results of the power of science and of concerns about the interference of funding agencies and industries has been a growing movement to deny the results of scientific research. Denialism is based on the idea that only 100 per cent agreement of scientists on a scientific topic is good enough to justify social action. This conflates two ideas that are commonly held by the general public. The first is that in an argument, both sides should be heard and given equal coverage. The second is that in the past minority views in science have been shown to be correct.

Deniers often portray themselves as being like Galileo standing against the Roman Catholic Church or Wegener challenging the authority of geologists. The problem is that these people are usually confusing scientific debate over details with lack of consensus on the overarching theory. Additionally, deniers use the idea that science is part of a larger social context to suggest that any particular idea is contingent and therefore not to be trusted. A good example of this is the modern flat-earth community. Although there is a wide variety of reasons why people come to believe the Earth is flat rather than a globe, to maintain that belief means that no amount of evidence contradicting their belief has any meaning to them. In many ways, modern flat-earthers are like people stuck in Thomas Kuhn's "pre-science" phase of scientific development. There is no central body of tested knowledge (no map with correct distances or measurement of the distance to the Sun), each believer has their own set of beliefs (for example, some believe we live on a disc moving through space, others believe we live on a stationary infinite plane), and there is a sense that there is some great secret involved, either in the form of a global conspiracy to hide the truth or that nature works by a different set of rules than those used to demonstrate the spherical earth. While a belief in a flat earth is largely harmless in itself, combined with other forms of anti-science reveals that a significant group of people are saying that the world is too much for them – too complex, changing too quickly and too uncertain. They want simplicity and certainty. They want their senses to be the true arbiter of reality, not some physics textbook with its mysterious and invisible particles and forces or the even crazier realm of quantum physics.

Climate Change

On the global level we can look at climate change as the greatest demonstration of the fight over the place of science in the wider society. Some aspects of climate change denial are similar to the fight over the link between smoking and cancer or the existence of acid rain. In both cases, large commercial interests were under attack by the mainstream scientific community and in response those interests funded or hired scientists to counter or simply deny the scientific evidence. When that failed, they sought to obscure the public reception of the evidence by arguing that more study was needed or to delay action by keeping the controversy going for as long as possible. Lobbying government officials and bringing court cases to try to overturn laws and regulations often became a testing ground for what would have more authority: economic power or scientific knowledge. The minority scientists who worked for the large corporate interests have often portrayed themselves as crusaders for truth and compare themselves to Galileo, Einstein, or Wegner, but unlike Galileo, Einstein, or Wegener, they were not making any new discoveries, only attacking the discoveries of others. It is perfectly reasonable to test the ideas of those who said a large portion of lung cancers were caused by smoking or that acid rain was caused by sulfur dioxide and nitrogen oxides in air pollution. However, failing to generate actual scientific evidence is often a signal that contrary opinions are sliding toward denialism rather than true scientific debate.

In 2005 Hurricane Katrina crashed into the coast of the United States leaving behind more than 1,800 dead and doing billions of dollars of damage. Was the hurricane just the storm of the century, or was it part of global climate change? Record temperatures, receding glaciers, and concerns about changing ocean currents have all been pointed to as part of a much larger change in global climate. The science of climate change is complex, and over the years it has become a battleground for various interest groups, with one side arguing that it is human created and potentially catastrophic, especially if we do not take significant measures to limit greenhouse gases as soon as possible. The other side argues that the climate is not really being affected by human activity, which is minor compared to natural forces, and that radical activity will stifle the economy and do nothing but waste time and money. Each side has claimed that science supports its position. By 2015 the scientific community had reached a general consensus, with over 99 per cent of climate researchers agreeing that the major factors in climate change are anthropogenic (human made). Although truth is not determined by consensus, what convinced climate scientists was not the opinions of other people, but their own work. The consensus follows the evidence, not the other way around. While there are still strong advocates against this position, it has become more difficult to find credible scientists willing to take the side of the climate change deniers. On the other hand, what to do about climate change is proving just as difficult as persuading governments and industry that it exists.

The prevailing scientific position has been articulated by the Intergovernmental Panel on Climate Change (IPCC). This panel, created by the World Meteorological Organization and the UN Environment Program in 1988, produced a series of four assessment reports from 1990 to 2007. These assessments became the basis for the UN Framework Convention on Climate Change, which in 1997 produced the Kyoto Protocol, a plan to reduce greenhouse gases.

The IPCC received the Nobel Peace Prize, jointly with Al Gore, in 2007. In 2014 the IPCC released its fifth report and a sixth report is scheduled for 2022. What is interesting about the IPCC is that it is not a body that conducts scientific research. Rather, it gathers all the scientific research previously conducted and decides, through democratic processes, what the current state of scientific knowledge is.

One hundred and ninety-nine states agreed to follow the Kyoto Protocol by 2010, although Japan and Russia have said they would not set new targets and Canada formally withdrew from the Protocol in 2012. The UN Climate Change Conference of 2015 (COP 21) in Paris resulted in a new agreement to take stronger measures to reduce greenhouse gas emissions, in order to keep global warming below 2°C. Despite these actions, the average temperature of the Earth continues to rise, and the ability of states and industry to turn this around will tax both science and governments. The history of this conflict reveals the difficulty of applying science to problems with political and economic implications. Developing countries don't want to stop increasing their economic activity and point out that on a per-person basis they pollute far less than people in developed countries. Industrial nations have resisted regulations that apply only to industry in developed countries, pointing out that industries in developing countries are often poorly regulated and use old technology. From a scientific point of view it may not matter who produces the pollution, but from a political point of view it certainly does. A number of countries refused to sign the agreement including Turkey, Iran, Iraq, Libya, and Yemen. A major blow to the global agreement was the rejection of the terms of the agreement by the United States in 2017 and its formal withdrawal in 2020. This demonstrates the need, as we move forward, for scientists and social scientists to work together with the public to shape government support and change social attitudes and behaviors as well as using technology to change industry.

COVID-19 and the Problem of Politics and Science

The importance of science was forcefully demonstrated in 2019 when the world learned about the appearance of a new form of coronavirus. Named COVID-19 (CO from corona, VI from virus, D for disease, and 19 for the year it was identified [2019]), the virus created a pandemic that spread across the globe in a few months. Although its mortality rate will take several years to accurately measure, most industrial nations had death rates between 2 per cent and 4 per cent of those infected. This created havoc as borders closed, businesses stopped operating, and people were urged or ordered to stay home.

Pandemics are not new – notable historical pandemics include the Black Death (1347), when the mortality rate was as high as 66 per cent in some places, and the influenza pandemic of 1918, when between 3 per cent and 5 per cent of the population of the world died. The difference with COVID-19 was the ability of the scientific community to mobilize to analyze the virus and offer advice for preventing its further spread. Within a few weeks of its appearance, there were massive amounts of information about the virus ranging from electron microscope images to maps of the genetic material of the virus. At the time of writing, more than a dozen vaccines are in trials and treatment protocols have lowered the mortality rate significantly.

Despite the major effort by scientists and medical researchers to defend against the spread of COVID-19 while the race for treatments was on, the impact of the pandemic was very different depending on where in the world you lived. In countries with strong medical systems and governments that listened to the advice of epidemiologists and other scientists, the impact of the disease was less than in countries that had poor medical systems or governments that refused to follow advice from experts. Thus, on October 20, 2020, the United States, with a poorly handled and fragmented response to COVID, had 25,453 cases per million people and 224,027 deaths, while New Zealand had a rigorous response with 384 cases per million people and only 25 deaths. While a small island nation has some natural advantages over the United States, another island nation, the United Kingdom, had 11,627 cases per million people and 45,712 deaths.[2] The United Kingdom's medical system is good, but the government response was slow and politicians were frequently at odds with their own experts. Countries that followed scientific advice have done better than comparable countries where politicians ignored or downplayed expert advice. It will be interesting to see if this fact makes politicians and the public more attentive to science in the future.

Pseudoscience

While the case of cold fusion demonstrates the dangers of commercial pressures and the fight over climate change represents the problem of the political utility of

---

2. Coronavirus Dashboard (Live), "COVID-19 Real-Time Data," https://covidly.com/.

science, our scientific age has also produced a rise in efforts to deceive people by using ideas that sound like science, but are in fact wrong or even fraudulent. The term "pseudoscience" refers to claims that something is scientific but lacks actual scientific evidence. The major problem with pseudoscience is that the use or application of such ideas or products uses up resources and can even endanger people. In some cases, the historical foundation of pseudoscientific beliefs had some justification such as vernalization used by Lysenko or the idea that personality could be determined by the shape of the skull in phrenology. In these cases what started as a reasonable hypothesis was demonstrated to be false, a process that happens all the time in science. They became pseudoscience because significant numbers of people continued to believe in them despite actual evidence to the contrary.

The most notorious pseudoscience in Western history has been homeopathy. The field of homeopathy was created in 1796 by Samuel Hahnemann (1755–1843) in an era before the germ theory of disease was discovered and bleeding was still a common practice among physicians. Hahnemann's impulse was a good one – do no harm – but he created a system based on pure fantasy. At its heart, homeopathic theory is based on the idea that "like cures like," so that if, for example, you have a fever, you should take something that would make you feel hot such as chili peppers.

Added to the "like cures like" dictum was the even more fanciful idea that the lower the dose of the curative, the greater its curative strength. This is achieved by serial dilution of one part to one hundred parts of solvent (1C), and then repeated. Hahnemann often used 30C dilutions (1 molecule of active ingredient to 10^{60} molecules of solute). In molecular terms, a 13C homeopathic remedy does not contain a single molecule of the original material. Dilutions up to 200C are used in modern homeopathic materials. To understand the scale of this, 200C is $1:10^{400}$ while the estimated number of atoms in the observable universe is only 10^{80}. In other words, most homeopathic remedies contain nothing but a small amount of solvent, usually distilled water. Even if the original material had some medical utility, it isn't in the final product.

The immunologist Jacques Benveniste (1935–2004) tried to rescue homeopathy with claims that water had memory, so that dilution to no original material was irrelevant. Because Benveniste, like Pons and Fleischmann, was a respected scientist, his ideas were scientifically tested even though they defied basic principles of science and even logic (why, for example, would the water remember the homeopathic material and not everything else it had been in contact with?).

Benveniste went even further in 1997 when he argued that the memory could be transmitted over telephone lines and later over the internet. A small number of people claimed to have replicated Benveniste's work, but in every case they failed to replicate the work when observed by third parties.

Homeopathy has depended for its continued existence on the support of celebrities and the fear of many people about the problems with the medical system. While it was perfectly reasonable for scientists to investigate homeopathic remedies, it is also perfectly reasonable to say that such remedies have no utility. They do, however, pose a threat to people who may not seek real medical help and they may also put others at risk by the use of homeopathic "vaccinations" that open a path for infectious diseases that are easily controlled to reoccur.

Conclusion

As science has had a greater and greater role in shaping society, there has always been resistance to the changes it offers. At one level, it is wise to be cautious about the introduction of new products in a complex system. As the critic of technology Neil Postman (1931–2003) pointed out, the introduction of a new "thing" – be it a device, practice, or ideology – changes society. It is not "society plus the computer," but a new society.[3] The conundrum of science in society is that the producers of science may not be the best people to judge what the impact of their products will be; however, because of the highly technical nature of the work, those who lack training may not understand the work well enough to make informed choices. Errors, flawed work, and other problems are inevitable, and the examples of DDT, thalidomide, and eugenics should stand as a warning that scientific mistakes can have dangerous consequences. The fact that there will be problems does not, however, mean that science should be rejected. Rather, it means that we must work to understand the potential benefits and problems of using scientific developments.

Scientific research represents a complex interplay of social demands, technical constraints, and personal interests and abilities. It is not driven solely by ideas, but neither can it be produced to order. While science has provided some profound insights into the structure of nature, it has also presented us with some

3. Neil Postman, *Technopoly* (New York: Vintage, 1993).

13.3 CMS HIGGS EVENT

Source: CERN. http://cdsweb.cern.ch/record/628469. Licensed under the terms
of CC-BY-SA (https://creativecommons.org/licenses/by-sa/4.0/).

difficult questions about how to use that knowledge. Ironically, knowing more has
made our choices more difficult rather than less. Understanding the history of
science offers another venue for approaching these difficult questions, since it can
show us the power and the danger of past choices and explain how we have
arrived at the world we live in. Science has, for example, been claimed as the
basis for both Marxism and modern democracy.

To complicate things even more, science today has a much more public face.
In the past, the average person might read about the Nobel prizes or see reports
about possible cures for cancer. Now, almost every important scientific event has a
website. When the physicists at CERN (*Conseil Européen pour la Recherche Nucléaire*)
announced the likely discovery of the Higgs boson particle (see figure 13.3) in 2013
and confirmed their work in 2017, the public could go to the CERN homepage and

find out what was happening. Anyone with a computer can help do research on a variety of projects including protein folding and the search for extra-terrestrial life. While public access is generally positive, it can also lead to information overload and unrealistic expectations.

The history of science can be useful because it reveals the broader context of science rather than looking only at its products. No one owns science. If we wish to make informed choices, we must never forget that science exists because people created it, and it cannot exist separate from the community. Behind all the patents, prizes, and professional degrees, the idea of science – our long effort to understand nature – and the knowledge that radiates from that search are part of our shared human heritage.

Essay Questions

1. What issues about science does the case of cold fusion expose?

2. What are common characteristics of pseudoscience?

3. Why did climate change deniers work against the scientific consensus?

4. How does corporate scientific research threaten the ideology of public knowledge?

5. In what ways can Indigenous knowledge and scientific knowledge work together to solve important problems?

CODA: THE PATH TO A GLOBAL HISTORY OF SCIENCE

As we have traced the messy and situated history of modern science, we are still left with an origin story that has deep European roots. While we want to see the product of science as universal in its application, its history remains very Eurocentric. Historians telling of the development of modern science have often erased the complex interactions between different groups of peoples and other knowledge systems and instead traced a sort of inevitability of science, developing out of its own internal logic. We might remember Alexis Guignard de Saint-Priest's dictum that "the history is right perhaps, but let us not forget, it was written by the victors."[1] This book has begun to challenge this interpretation and see the interconnections through time and space of different knowledge systems in conversation and in conflict. Modern science is a hybrid of different knowledge systems, rather than the triumph of the European model over the rest of the world. This is a continuing dialogue.

History, like science, evolves as new information, new voices, and new conversations shape the discipline. We challenge historians who come after us to write a history of science that foregrounds these conversations, a sort of cosmopolitan history of science, that would place the rich interconnections of various

---

1. Alexis Guignard de Saint-Priest, "... *l'histoire est juste peut-être, mais qu'on ne l'oublie pas, elle a été écrite par les vainqueurs,*" *Histoire de la royauté considérée dans ses origines, jusqu'à la formation des principales monarchies de l'Europe*, vol. 2 (Paris: H.I. Delloye, 1842), 42.

intellectual traditions over the years at the center of the story. If we think about a nonhierarchical interaction of knowledge communities, this could lead to a richer way to understand both the development of science around the world, and the state of science today, particularly in dialogue with traditional knowledge systems.

FURTHER READING

FIVE   THE SCIENTIFIC REVOLUTION: CONTESTED TERRITORY

Bala, Arun, ed. *Asia, Europe and the Emergence of Modern Science: Knowledge Crossing Boundaries*. New York: Palgrave, 2012.

Dear, Peter Robert. *Revolutionizing the Sciences: European Knowledge and Its Ambitions, 1500–1700*. Princeton, NJ: Princeton University Press, 2001.

Harkness, Deborah. *The Jewel House: Elizabethan London and the Scientific Revolution*. New Haven, CT: Yale University Press, 2007.

Hunter, Lynette, and Sarah Hutton, eds. *Women, Science and Medicine 1500–1700: Mothers and Sisters of the Royal Society*. Stroud, UK: Sutton, 1997.

Jardine, Lisa. *Ingenious Pursuits: Building the Scientific Revolution*. New York: Nan A. Talese, 1999.

Lindberg, David C., and Robert S. Westman, eds. *Reappraisals of the Scientific Revolution*. Cambridge: Cambridge University Press, 1990.

Long, Pamela O. *Artisans/Practitioners and the Rise of the New Sciences, 1400–1600*. Corvallis: Oregon State University Press, 2011.

Newman, William Royall. *Atoms and Alchemy: Chymistry and the Experimental Origins of the Scientific Revolution*. Chicago: University of Chicago Press, 2006.

Newton, Isaac. *Opticks*. New York: Prometheus, 2003.

———. *The Principia*. Translated by Andrew Motte. New York: Prometheus, 1995.

Osler, Margaret J. *Reconfiguring the World: Nature, God, and Human Understanding from the Middle Ages to Early Modern Europe*. Baltimore, MD: Johns Hopkins University Press, 2010.

———. *Rethinking the Scientific Revolution*. Cambridge: Cambridge University Press, 2000.

Park, Katherine, and Lorraine Daston, eds. *Early Modern Science. Vol. 3 of The Cambridge History of Science*. Cambridge: Cambridge University Press, 2003.

Schiebinger, Londa L. *The Mind Has No Sex?: Women in the Origins of Modern Science*. Cambridge, MA: Harvard University Press, 1989.

Shapin, Steven. *The Scientific Revolution*. Chicago: University of Chicago Press, 1996.

Smith, Pamela H. *The Body of the Artisan: Art and Experience in the Scientific Revolution*. Chicago: University of Chicago Press, 2004.

Westfall, Richard S. *Never at Rest: A Biography of Isaac Newton*. Cambridge: Cambridge University Press, 1980.

SIX   THE ENLIGHTENMENT AND ENTERPRISE

Bell, Madison Smartt. *Lavoisier in the Year One: The Birth of a New Science in an Age of Revolution*. New York: W.W. Norton, 2006.

Binnema, Ted. *Enlightened Zeal: The Hudson's Bay Company and Scientific Networks, 1670–1870*. Toronto: University of Toronto Press, 2014.

Diderot, Denis. *Encyclopédie ou dictionnaire raisonné des sciences des arts et des métiers*. Stuttgart-Bad Cannstatt: F. Frommann Verlag (G. Holzboog), 1966.

Fox, Christopher, Roy Porter, and Robert Wokler, eds. *Inventing Human Science: Eighteenth-Century Domains*. Berkeley: University of California Press, 1995.

Gascoigne, John. *Encountering the Pacific in the Age of Enlightenment*. Cambridge: Cambridge University Press, 2014.

Holmes, Frederic Lawrence. *Lavoisier and the Chemistry of Life*. Madison: University of Wisconsin Press, 1985.

Lavoisier, Antoine. *Elements of Chemistry*. Translated by Robert Kerr. New York: Dover, 1965.

Lynn, Michael R. *Popular Science and Public Opinion in Eighteenth-Century France*. Manchester: Manchester University Press, 2006.

Poirier, Jean-Pierre. *Lavoisier, Chemist, Biologist, Economist*. Translated by Rebecca Balinski. Philadelphia: University of Pennsylvania Press, 1996.

Porter, Roy, ed. *Eighteenth-Century Science. Vol. 4 of The Cambridge History of Science*. Cambridge: Cambridge University Press, 2003.

Schiebinger, Londa L. *Colonial Botany: Science, Commerce, and Politics in the Early Modern World*. Philadelphia: University of Pennsylvania Press, 2005.

SEVEN   SCIENCE AND EMPIRE

Bartholomew, James. *The Formation of Science in Japan: Building a Research Tradition*. New Haven, CT: Yale University Press, 1989.

Bowler, Peter J. *Evolution: The History of an Idea*. Berkeley: University of California Press, 1984.

———. *Life's Splendid Drama: Evolutionary Biology and the Reconstruction of Life's Ancestry, 1860–1940*. Chicago: University of Chicago Press, 1996.

Darwin, Charles. *The Origin of Species by Means of Natural Selection, or, The Preservation of Favoured Races in the Struggle for Life*. London: J. Murray, 1860.

Garfield, Simon. *Mauve: How One Man Invented a Color That Changed the World*. New York: Norton, 2001.

Geison, Gerald L. *The Private Science of Louis Pasteur*. Princeton, NJ: Princeton University Press, 1995.

Greene, Mott T. *Geology in the Nineteenth Century: Changing Views of a Changing World*. Ithaca, NY: Cornell University Press, 1982.

Lyell, Charles. *Principles of Geology*. Chicago: University of Chicago Press, 1990.

Nye, Mary Jo. *Before Big Science: The Pursuit of Modern Chemistry and Physics, 1800–1940*. New York: Twayne Publishers, 1996.

Paul, Harry W. *From Knowledge to Power: The Rise of the Science Empire in France, 1860–1939*. Cambridge: Cambridge University Press, 1985.

Strathern, Paul. *Mendeleyev's Dream: The Quest for the Elements*. New York: St. Martin's Press, 2000.

EIGHT   ENTERING THE ATOMIC AGE

Dry, Sarah. *Curie*. London: Haus Publishers, 2003.

Ede, Andrew. *The Rise and Decline of Colloid Science in North America, 1900–1935: The Neglected Dimension*. Aldershot, UK: Ashgate, 2007.

Hunt, Bruce J. *Pursuing Power and Light: Technology and Physics from James Watt to Albert Einstein*. Baltimore, MD: Johns Hopkins University Press, 2010.

Knight, David. *The Making of Modern Science: Science, Technology, Medicine and Modernity, 1789–1914*. Cambridge: Polity Press, 2009.

Levinovitz, Agneta Wallin. *The Nobel Prize: The First 100 Years*. London: Imperial College Press, 2001.

Navarro, James. *A History of the Electron: J.J. and G.P. Thomson*. Cambridge: Cambridge University Press, 2012.

Nye, Mary Jo, ed. *The Modern Physical and Mathematical Sciences. Vol. 5 of The Cambridge History of Science*. Cambridge: Cambridge University Press, 2003.

Pasachoff, Naomi E. *Marie Curie and the Science of Radioactivity*. New York: Oxford University Press, 1997.

Reeves, Richard. *A Force of Nature: The Frontier Genius of Ernest Rutherford*. New York: W.W. Norton, 2008.

Wilson, David. *Rutherford, Simple Genius*. Cambridge, MA: MIT Press, 1983.

NINE   SCIENCE AND WAR

Freemantle, Michael. *The Chemists' War, 1914–1918*. Cambridge: Royal Society of Chemistry, 2015.

Henig, Robin Marantz. *The Monk in the Garden: The Lost and Found Genius of Gregor Mendel, the Father of Genetics*. Boston: Mariner Books, 2001.

Isaacson, Walter. *Einstein: His Life and Universe*. New York: Simon & Schuster, 2008.

Mawer, Simon. *Gregor Mendel: Planting the Seeds of Genetics*. New York: Abrams, in association with the Field Museum, Chicago, 2006.

Russell, Edmund. *War and Nature: Fighting Humans and Insects with Chemicals from World War I to* Silent Spring. Cambridge: Cambridge University Press, 2001.

Shapiro, Adam R. *Trying Biology: The Scopes Trial, Textbooks, and the Antievolution Movement in American Schools*. London: Pickering and Chatto, 2013.

Stachel, John, ed. *Einstein's Miraculous Year: Five Papers That Changed the Face of Physics*. Princeton, NJ: Princeton University Press, 1998.

van Dongen, Jeroen. *Einstein's Unification*. Cambridge: Cambridge University Press, 2010.

TEN   THE DEATH OF CERTAINTY

Brennan, Richard P. *Heisenberg Probably Slept Here: The Lives, Times, and Ideas of the Great Physicists of the 20th Century*. New York: Wiley, 1997.

Cassidy, David Charles. *Uncertainty: The Life and Science of Werner Heisenberg*. New York: W.H. Freeman, 1992.

Galison, Peter. *How Experiments End*. Chicago: University of Chicago Press, 1987.

Galison, Peter, and Bruce Hevly, eds. *Big Science: The Growth of Large-Scale Research*. Stanford, CA: Stanford University Press, 1992.

Gimbel, Steven. *Einstein's Jewish Science: Physics at the Intersection of Politics and Religion*. Baltimore, MD: Johns Hopkins University Press, 2012.

Hughes, Jeff. *The Manhattan Project: Big Science and the Atom Bomb*. Cambridge: Icon Books, 2002.

Kohler, Robert E. *Partners in Science: Foundations and Natural Scientists*. Chicago: University of Chicago Press, 1991.

Lindee, M. Susan. *Suffering Made Real: American Science and the Survivors at Hiroshima*. Chicago: University of Chicago Press, 1994.

Maddox, Robert James. *Weapons for Victory: The Hiroshima Decision Fifty Years Later*. Columbia, MO: University of Missouri Press, 1995.

Olby, Robert C. *The Path to the Double Helix: The Discovery of DNA*. Foreword by Francis Crick. New York: Dover Publications, 1994.

Sime, Ruth Lewin. *Lise Meitner: A Life in Physics*. Berkeley: University of California Press, 1996.

Watson, James D. *The Double Helix: A Personal Account of the Discovery of the Structure of DNA*. London: Weidenfeld and Nicolson, 1997.

ELEVEN   1957: THE YEAR THE WORLD BECAME A PLANET

Brzezinski, Matthew. *Red Moon Rising: Sputnik and the Hidden Rivalries That Ignited the Space Age*. New York: Times Books, 2007.

Dickson, Paul. *Sputnik: The Launch of the Space Race*. Toronto: Macfarlane, Walter and Ross, 2001.

Launius, Roger D., and Howard E. McCurdy, eds. *NASA Spaceflight: A History of Innovation*. Cham, Switzerland: Palgrave Macmillan, 2018.

LeGrand, H. E., and Naomi Oreskes. *Plate Tectonics: An Insider's History of the Modern Theory of the Earth*. Boulder, CO: Westview, 2003.

Marvin, Ursula B. *Continental Drift: The Evolution of a Concept*. Washington, DC: Smithsonian Institution Press, 1973.

NASA. *Space Flight: The First 30 Years*. Washington, DC: National Aeronautics and Space Administration, Office of Space Flight, 1991.

Oreskes, Naomi. *The Rejection of Continental Drift: Theory and Method in American Earth Science*. New York: Oxford University Press, 1999.

Stine, G. Harry. *ICBM: The Making of the Weapon That Changed the World*. New York: Orion Books, 1991.

Sullivan, Woodruff Turner. *Cosmic Noise: A History of Early Radio Astronomy*. Cambridge, UK: Cambridge University Press, 2009.

Wang, Zuoyue. *In Sputnik's Shadow: The President's Science Advisory Committee and Cold War America*. New Brunswick, NJ: Rutgers University Press, 2008.

Wegener, Alfred. *The Origin of Continents and Oceans*. Translated by John Biram. London: Methuen, 1966.

Wilson, J. Tuzo. *IGY: The Year of the New Moons*. London: Michael Joseph, 1961.

TWELVE MAN ON THE MOON, MICROWAVE IN THE KITCHEN

Agar, Jon. *Turing and the Universal Machine: The Making of the Modern Computer*. Cambridge: Icon, 2001.

Alic, Margaret. *Hypatia's Heritage: A History of Women in Science from Antiquity to the Late Nineteenth Century*. London: Women's Press, 1986.

Aspray, William. *John von Neumann and the Origins of Modern Computing*. Cambridge, MA: MIT Press, 1990.

Campbell-Kelly, Martin, William Aspray, Nathan Ensmenger, and Jeffery R. Yost. *Computer: A History of the Information Machine*. 3rd ed. Boulder, CO: Westview Press, 2014.

Carson, Rachel. *Silent Spring*. Introduction by Al Gore. Boston: Houghton Mifflin, 1994.

Clancey, William J. *Working on Mars: Voyages of Scientific Discovery with the Mars Exploration Rovers*. Cambridge, MA: MIT Press, 2012.

Coles, Peter. *Cosmology: The Origin and Evolution of Cosmic Structure*. Chichester, UK: John Wiley, 2002.

Corrigan, Jim. *Alan Turing*. Greensboro, N.C.: Morgan Reynolds Pub., 2009.

Des Jardins, Julie. *The Madame Curie Complex: The Hidden History of Women In Science*. New York: Feminist Press at City University of New York, 2010.

Hodges, Andrew. *Turing: A Natural Philosopher*. London: Phoenix, 1997.

McElheny, Victor K. *Drawing the Map of Life: Inside the Human Genome Project*. New York: Basic Books, 2010.

McLaren, Angus. *A History of Contraception: From Antiquity to the Present Day*. Oxford: Basil Blackwell, 1990.

Riordan, Michael. *The Hunting of the Quark: A True Story of Modern Physics*. New York: Simon & Schuster, 1987.

Rossiter, Margaret. *Women Scientists in America: Before Affirmative Action, 1940–1972*. Baltimore, MD: Johns Hopkins University Press, 1995.

Sideris, Lisa H. *Rachel Carson: Legacy and Challenge*. Albany, NY: State University of New York Press, 2008.

Smith, Robert W. *The Space Telescope: A Study of NASA, Science, Technology and Politics*. 2nd ed. Cambridge: Cambridge University Press, 1993.

Stein, Dorothy. *Ada, A Life and a Legacy*. Cambridge, MA: MIT Press, 1985.

Swade, Doron. *The Difference Engine: Charles Babbage and the Quest to Build the First Computer*. New York: Viking, 2001.

THIRTEEN SCIENCE AND NEW FRONTIERS: POTENTIAL AND PERIL IN THE NEW MILLENNIUM

Aldersey-Williams, Hugh. *The Most Beautiful Molecule: The Discovery of the Buckyball*. New York: Wiley, 1995.

Clarke, Arthur Charles. *Ascent to Orbit: A Scientific Autobiography: The Technical Writings of Arthur C. Clarke*. New York: Wiley, 1984.

Cowan, Ruth Schwartz. *Heredity and Hope: The Case for Genetic Screening*. Cambridge, MA: Harvard University Press, 2008.

Drexler, K. Eric. *Engines of Creation: The Coming Era of Nanotechnology*. London: Fourth Estate, 1996.

Mulhall, Douglas. *Our Molecular Future: How Nanotechnology, Robotics, Genetics, and Artificial Intelligence Will Transform Our World*. Amherst, NY: Prometheus Books, 2002.

Oreskes, Naomi, and Eric M. Conway. *Merchants of Doubt: How a Handful of Scientists Obscured the Truth on Issues from Tobacco Smoke to Global Warming*. New York: Bloomsbury Press, 2010.

Rifkin, Jeremy. *The Biotech Century: Harnessing the Gene and Remaking the World*. New York: Jeremy P. Tarcher/Putnam, 1998.

Taubes, Gary. *Bad Science: The Short Life and Weird Times of Cold Fusion*. New York: Random House, 1993.

GENERAL READINGS IN THE HISTORY OF SCIENCE

Bowler, Peter J., and Iwan Rhys Morus. *Making Modern Science: A Historical Survey*. Chicago: University of Chicago Press, 2005.

Brock, William H. *The Chemical Tree: A History of Chemistry*. New York: Norton, 2000.

———. *The Norton History of Chemistry*. New York: Norton, 1992.

Bronowski, Jacob. *The Ascent of Man*. Boston: Little, Brown, 1973.

Bryson, Bill. *A Short History of Nearly Everything*. New York: Broadway Books, 2003.

Diamond, Jared. *Guns, Germs, and Steel: The Fates of Human Societies*. New York: W.W. Norton, 2005.

Golinski, Jan. *Making Natural Knowledge: Constructivism and the History of Science*. Cambridge: Cambridge University Press, 1998.

North, John David. *The Norton History of Astronomy and Cosmology*. New York: Norton, 1995.

Olby, Robert C. *Fontana History of Biology*. New York: Fontana, 2002.

Taub, Liba Chaia, Alexander Jones, et al. *The Cambridge History of Science*. 7 vols. Cambridge: Cambridge University Press, 2003.

INDEX

Illustrations indicated by page numbers in italics